Bose Einstein Condensation of Excitons and Polaritons

Authored by

Sunipa Som

Nehru Gram Bharati University, Prayagraj-221002, India

Bose Einstein Condensation of Excitons and Polaritons

Author: Sunipa Som

ISBN (Online): 978-981-5165-40-1

ISBN (Print): 978-981-5165-41-8

ISBN (Paperback): 978-981-5165-42-5

Published by Bentham Science Publishers Pte. Ltd. Singapore. All Rights Reserved.

First published in 2024.

need for a court order if at any point you breach any terms of this License Agreement. In no event will any delay or failure by Bentham Science Publishers in enforcing your compliance with this License Agreement constitute a waiver of any of its rights.

3. You acknowledge that you have read this License Agreement, and agree to be bound by its terms and conditions. To the extent that any other terms and conditions presented on any website of Bentham Science Publishers conflict with, or are inconsistent with, the terms and conditions set out in this License Agreement, you acknowledge that the terms and conditions set out in this License Agreement shall prevail.

Bentham Science Publishers Pte. Ltd.
80 Robinson Road #02-00
Singapore 068898
Singapore
Email: subscriptions@benthamscience.net

CONTENTS

PREFACE

The possibility of Bose Einstein condensation (BEC) of excitons in semiconductors has been an interesting topic for many years. BEC is the macroscopic occupation of the lowest energy quantum state of a system of bosons. This state of matter is very difficult to observe but it is a notable example of quantum mechanics on a macroscopic scale. First the concept of Bose condensation was given by Satyendra Nath Bose and Albert Einstein in 1924. After seventy years, BEC was demonstrated for the first time in a dilute gas of rubidium atoms in 1995.

Exciton is an electrically neutral quasiparticle. Due to the Coulomb force, an electron and a hole are attracted to each other and form a bound state known as an exciton. It can transport energy without transporting charge. Exciton exists in insulators, semiconductors, and some liquids. These quasiparticles are the natural candidates for observing the BEC due to its boson nature and its light-effective mass.

Exciton–polaritons (or polaritons for short) are bosonic quasiparticles produced as a result of the coupling between excitons and the electromagnetic field. When a cavity photon and an exciton are superimposed, a polariton is formed that exists inside semiconductor microcavities. Due to its bosonic nature and very light effective mass, typically of the order of $10-4$ times of the bare electron mass, above a critical density, the polaritons macroscopically occupy its lowest quantum state and form a condensate. The polaritons have a very short lifetime and due to this, it is inherently non-equilibrium in nature. However, many features of the polaritons are similar to the features that we can expect for the equilibrium BEC. These natures of polariton make it an interesting topic not only from a fundamental point of view but also from its potential practical application in future quantum technological devices.

The research on the BEC of excitons and polaritons is very important for the field of quantum information processing. BEC can be used to make quantum computers and is useful as a quantum simulator. BEC of excitons and polariton is a rapidly developing field of solid-state physics that motivated me to write this book.

This book offers an overview of the BEC of excitons and BEC of polaritons. It contains the history of BEC (Chap. 1), fundamentals of excitons (Chap. 2), fundamentals of polaritons (Chap. 3), BEC of excitons (Chap. 4), and BEC of polaritons (Chap. 5). The origin of excitons, relaxation and thermalization behaviour of Excitons have been discussed in Chapter 2. Chapter 3 contains information about the origin of polaritons, special condensate features like polariton lasing, super-fluidity, and quantized vortices. Experimental techniques, theoretical modeling, recent developments, and application of BEC of excitons and polaritons have been discussed in Chapters 4 and 5, respectively.

This book puts together the fundamentals of excitons and polaritons, all the advances, recent research works, discoveries, and applications on the BEC of excitons and polaritons. It also gives a brief outlook on future work. I believe that its straight-forward content will make it accessible and interesting to a broad readership, ranging from students, research fellows, lecturers, and engineering professionals in the interdisciplinary domains of semiconductor physics, nanotechnology, photonics, materials science, and quantum physics.

Sunipa Som
Nehru Gram Bharati University, Prayagraj-221002,
India

ACKNOWLEDGEMENTS

Every book is a project that takes a long time and is the result of hard work, passion, and cooperation from other people. Therefore, I would like to express my gratitude to all the people who helped and encouraged me to write this book.

I am grateful to the publishing team of Bentham Science and specially Mrs. Fariya Zulfiqar (Manager of Publication) for their continuous encouragement, support, and patience.

Similarly, I am grateful to Prof. Heinrich Stolz, University of Rostock, Germany who supervised me for a few years and helped me gain interest and knowledge in the field of exciton, and polariton Physics. I am also grateful to my former co-workers at the University of Rostock, Germany, specially Dr. Frank Kieseling and Dr. Rico Scwratze for their useful discussions.

Special thanks to Prof. Ram Kripal, University of Allahabad, India for his continuous encouragement. I also would like thank to my colleagues at the Nehru Gram Bharati University, Prayagraj, India.

Similarly, I am grateful to all the researchers and authors in the field of exciton polariton Physics whose works helped me to get a lot of information and write this book.

Last but not least, I am grateful for the continuous support and encouragement from my family.

CHAPTER 1

Introduction

Abstract: Bose Einstein Condensate (BEC) is the fifth state of matter in condensed matter physics. It is the most impressive example of quantum behaviour on a macroscopic scale but the most difficult state to observe. When low-density boson gas is cooled to a temperature nearly absolute zero, then BEC is formed. In this condition, the bosons have almost no free energy to move relative to each other. They clump together, come into the lowest quantum state, and behave as a single atom. Then microscopic quantum mechanical phenomena work on a macroscopic scale. This chapter summarizes the history and the developments regarding BEC with a brief discussion of the key scientists in this field. In the next section, the summary of the all chapters is given.

Keywords: Bose Einstein Condensate, History of Bose Einstein Condensation, Exciton, Polariton.

HISTORY OF BOSE EINSTEIN CONDENSATION

In nature, there are two types of particles, bosons and fermions, depending on their spin. The particle with integer spin is known as Boson. Bosons occupy the same quantum state because Bosons do not obey the Pauli Exclusion Principle. Whereas a particle with half-integer spin is known as a fermion. Fermions obey the Pauli Exclusion Principle and as a result, they cannot simultaneously occupy the same quantum state. Bosons can form BEC when their average separation becomes comparable to their thermal de Broglie wavelength. We know that the de Broglie wavelength $\lambda_D = \sqrt{\dfrac{2\pi\hbar^2}{mk_BT}}$ is inversely proportional to the square root of the temperature and particle mass. Therefore, the BEC of bosons is formed at a high temperature with light effective mass.

The first idea of BEC appeared in a paper about a new derivation of photon statistics and Plank distribution [1] which was written by Satyendra Nath Bose in 1924. In his paper, he derived Plank's quantum radiation law in a new way where he derived it on the basis of quantum statistics' but not on the basis of classical Mechanics. Then he sent this paper to Albert Einstein. Einstein was impressed by his work, he translated this paper into German language and submitted it in the Journal, Zeitschrift fur Physik in the name of Satyendra Nath Bose which was published in 1924. After that, Einstein extended this idea in the case of matter in two papers. He [2] gave the theoretical description of BEC for a homogeneous system of identical particles in 1925. In this theoretical description, he speculated that because of atoms condensed in its lowest energy state, the phase transition takes place in case of non-interacting atomic gas and it is the result of quantum statistical effect. From the results of Satyendra Nath Bose and Albert Einstein works, an idea of Bose Einstein statistics came out that describes the statistical distribution of Bosons. Boson particles include photons and atoms like helium-4.

After a long time, there was no noticeable work done on that. Then in 1937, it had been discovered that helium II is a superfluid. In January 1938, this work was published by a Soviet Physicist, Pyotr Kapitsa and also by Canadian Scientists John F Allen and Don Misener at the University of Toronto. Helium was first invented by Kammerlingh Onnes in 1908.

Satyendra Nath Bose

Satyendra Nath Bose was born on 1st January 1894 and died on 4th February 1974. He was an Indian theoretical physicist and mathematician. He is well known for his work on Bose Einstein condensate. In the early 1920, he gave the theoretical description of BEC. In 1954, he was awarded the Padma Vibhushan by the Government of India. He was also a Fellow of the Royal Society.

Albert Einstein

Albert Einstein, was born on March 14, 1879, Ulm, Württemberg, Germany and died on April 18, 1955, Princeton, New Jersey, U.S. He is well known for developing the special and general theories of relativity and in 1921 he won the Nobel Prize in physics for the explanation of the photoelectric effect. His mass energy equivalence relation $E=mc^2$, is the world's most popular equation.

Kammerlingh Onnes

Heike Kamerlingh Onnes a Dutch physicist and Nobel laureate was born on 21st September 1853 and died on 21st February 1926. He investigated the materials behaviour at or nearly absolute zero temperature. For this work, he used the Hampson Linde cycle and later in 1908, he used it to liquefy helium for the first time. He also discovered superconductivity in 1911.

Fritz London

Fritz Wolfgang London, a German physicist and professor at Duke University, was born on March 7, 1900 and died on March 30, 1954. He made basic and notable contributions to the theories of chemical bonding and intermolecular forces which is known as London dispersion forces. He and his brother Heinz London played a great role in understanding the electromagnetic properties of superconductors with the London equations. On five separate occasions, his name was recommended for the Nobel Prize in chemistry.

During the period of late 1920 to early 1930 it was observed that liquid helium II has some unusual properties. In 1938, Scientists Allen, Misener and Kapitza observed a frictionless flow of helium II that is called superfluid. The super fluidity of liquid helium-4 occurs at much higher temperatures than liquid helium-3. The atoms of helium-4 have an integer spin and therefore, these are bosons. The atoms of helium-3 are fermions. Bosons can form helium-3 by pairing one particle with another particle at much lower temperatures. This process is the same as the electron pairing in superconductivity. In 1996, the Nobel Prize in physics was given to David M. Lee, Douglas D. Osheroff, and Robert C. Richardson for the discovery of superfluidity in helium-3.

Nikolay Bogolyubov

Nikolay Nikolayevich Bogolyubov was a Soviet and Russian mathematician and physicist. He was born on 21st August 1909 and died on 13th February 1992. His contribution to classical and quantum statistical mechanics, quantum field theory and the theory of dynamical systems is notable and made him popular. He was awarded the Dirac Medal in 1992.

Lev Davidovich Landau

Lev Davidovich Landau was a Soviet theoretical physicist. He was born on Jan 22, 1908, Baku, Russian Empire (now Azerbaijan) and died on April 1, 1968, Moscow, Russia, U.S.S.R.. He is well known for his pioneering work on the quantum theory of condensed matter. He was awarded the Nobel Prize in physics in 1962.

Evgeny Mikhailovich Lifshitz

Evgeny Mikhailovich Lifshitz was a Soviet physicist. He was born on February 21, 1915, Kharkiv, Russian Empire and died on October 29, 1985, Moscow, Russian SFSR. He is famous in the field of general relativity.

Roger Penrose

Sir Roger Penrose was a mathematician, mathematical physicist, and a philosopher of science. He was born on 8th August 1931, Colchester, Essex, England. He was awarded a Nobel Prize in physics, in 2020 for his work on black hole. He is Emeritus Rouse Ball Professor of Mathematics in the University of Oxford, an emeritus fellow of Wadham College, Oxford, and an honorary fellow of St John's College, Cambridge and University College London.

Immediately after that in 1938, Fritz London claimed that BEC could be the cause of superfluidity of helium-4 and superconductivity. He also proposed that superfluidity and superconductivity are quantum processes on an observable scale.

But before the late 1950, the importance of this idea was not clear. In 1941, a Scientist Landau developed a theory of superfluids on the basis of the spectrum of elementary excitations of the fluid. In 1947, a Scientist Bogoliubov first established the theory of interacting Bose gasses from the concept of BEC. After that many theoretical works have been carried out to understand the relationship between BEC and superfluidity. To understand the relationship between superfluidity and BEC, intense theoretical work was developed. In 1951, first the concept of the non-diagonal long range order was given by Landau, Lifshitz and Penrose. Then in 1956, Penrose and Onsager discussed its relationship with BEC. At the same time, an experimental work on superfluid helium had become increased to check Landau's predictions.

Lars Onsager

Lars Onsager was born on November 27, 1903, Norway and died on October 5, 1976. He was an American physical chemist and theoretical physicist. He held the Gibbs Professorship of theoretical chemistry at Yale University. He was awarded the Nobel Prize in chemistry in 1968 for the discovery of Onsager reciprocal relations. Those relations are fundamental for the thermodynamics of irreversible processes.

Richard Phillips Feynman

Richard Phillips Feynman was born on 11th May 1918, New York, USA and died on 15th February 1988, Los Angeles, USA. He was an American theoretical physicist. He is well known for his work in the path integral formulation of quantum mechanics, the theory of quantum electrodynamics, and the physics of the superfluidity of super cooled liquid helium, as well as for his work in particle physics. He proposed the Parton model. For his contributions to the development of quantum electrodynamics, Feynman received the Nobel Prize in physics in 1965 jointly with Julian Schwinger and Shin'ichirō Tomonaga.

Henry Edgar Hall

Henry Edgar Hall, Fellow of the Royal Society (FRS), was born on 1928 and died on 4th December, 2015. He was a professor of low temperature physics at the University of Manchester. He was awarded the Guthrie Medal and Prize in 2004. He has important contributions to the research of superfluidity in both liquid ^{4} He and liquid ^{3} He. He worked on the function of topological defects, such as quantized vortex filaments in superfluid. He also contributed to the development of the ^{3} He/^{4} He dilution refrigerator that has a vital role in experimental work for the temperatures below 1 K.

William Frank Vinen

William Frank Vinen was born on 15th February 1930 and died on 8th June 2022. He was a British physicist and is well known for his work on low temperature physics. Vinen was elected as a Fellow of the Royal Society (FRS) in 1973. Henry Hall and Vinen's second-sound experiments in rotating helium first demonstrate the probable existence of vortex lines and later in 1950, the first indirect evidence of the quantization came from Vinen's helium heat currents experiments. He also investigated the flux flow and dissipation processes in type II superconductors.

First, the concept of quantized vortices was given by Onsager in 1949 and Feynman in 1955. Then it was experimentally proved by Hall and Vinen in 1956. This was an important development in this field.

The idea of the half-quantum vortex was given by the scientists G. E. Volovik and V. P. Mineev [3] and a few months later, M. C. Cross and W. F. Brinkman [4], in 1975. They proposed a new type of excitation known as the half-quantum vortex from the spinor nature of superfluids. A paper published in 1976 by two program directors, William Stwalley and Lewis Nosanow of the National Science Foundation [5] stimulated the investigation procedure to produce a Bose Einstein condensate. The four independent research groups immediately started working on this idea. These four groups were from the University of Amsterdam led by Isaac Silvera, the University of British Columbia led by Walter Hardy, Massachusetts Institute of Technology led by Thomas Greytak and Cornell University led by David Lee [6].

Eric Allin Cornell

Eric Allin Cornell was born on 19th December 1961. He is an American physicist. In 1995, he and Carl E. Wieman were able to produce the first Bose Einstein condensate. For their work, he was awarded the Nobel Prize in physics in 2001 with Wieman, and Wolfgang Ketterle.

Carl Edwin Wieman

Carl Edwin Wieman was born on March 26, 1951. He is an American physicist and presently the A.D White Professor at Large at Cornell University. He and Eric Allin Cornell create the first Bose Einstein condensate (BEC) in 1995. In 2001, Carl Wieman, Eric Cornell and Wolfgang Ketterle got the Nobel Prize in Physics for their studies on BEC.

Wolfgang Ketterle

Wolfgang Ketterle is a German physicist and professor of physics at the Massachusetts Institute of Technology (MIT). He was born on 21st October 1957, Heidelberg Gemany. His main experimental research work was on the trapped and cooled atoms for the temperatures close to absolute zero. He and his group members first realized the Bose Einstein condensation in 1995. In 2001, for this work and fundamental studies on condensates, he got the Nobel Prize in physics with Eric Allin Cornell and Carl Wieman.

The first gaseous condensate was discovered by Eric Cornell and Carl Wieman on 5th June 1995. It was produced in a gas of rubidium atoms cooled to 170 nanokelvin (nK) at the University of Colorado at Boulder NIST JILA lab [7]. After that, Wolfgang Ketterle produced a Bose Einstein Condensate in a gas of sodium atoms at MIT. In 2001, Cornell, Wieman, and Ketterle received the Nobel Prize in physics for their achievements [8]. In 1996, a Physicist J. R. Kirtley first observed the half quantum vortices in high-temperature superconductors [9]. A lot of experimental and theoretical works have been done on condensates of dilute atomic masses in the next few years. From the results of these works, a clear idea about the properties of equilibrium condensates was obtained.

Rainer Blatt

Rainer Blatt is a German-Austrian experimental physicist. He was born on 8th September 1952, Idar Oberstein, Germany. He is well known for his research on quantum optics and quantum information processing. The first experiments on teleport atoms were performed by his group, which was published in nature. He is a professor of physics at the University of Göttingen and a chair of experimental physics at the University of Innsbruck. Since 2003, he has been the Scientific Director at the Institute for Quantum Optics and Quantum Information (IQOQI) of the Austrian Academy of Sciences.

Igor Moskalenko

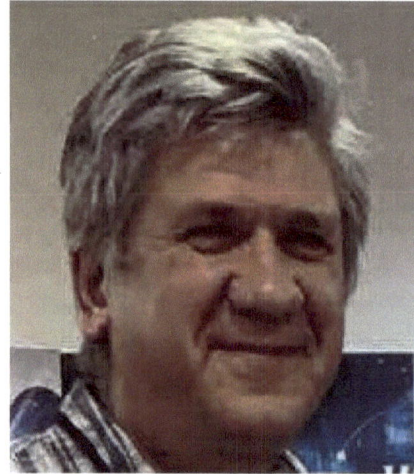

Igor Moskalenko was born on 4th May 1962, Moscow, Russia. He is a senior research scientist at the Hansen Experimental Physics Laboratory, Stanford University, Stanford, California, USA. He is a Fellow of American Physical Society.

The research development in the field of condensation stimulated the research in the field of semiconductor. In 1962, a Scientist Blatt and Moskalenko proposed that excitons can be a good candidate for observing the Bose Einstein condensate [10]. The electrons and holes bound together by the coulomb interaction and formed excitons. Due to the bosonic nature and light mass, excitons are a good candidate for the studies of Bose Einstein condensation. Though in 1931, a Scientist Yakov Frenkel first recommended the concept of exciton. He claimed that, without the net transfer of charge, the exciton can travel like particle through the lattice [11, 12].

After the discovery of exciton, scientists discovered a new quasiparticle polariton [13], which is a composite boson with a half-exciton and half-photon nature. The mass of polariton is four orders of magnitude lighter than the exciton and, therefore, it allows condensation even at room temperature theoretically.

Yakov Il'ich Frenkel

Yakov Il'ich Frenkel was a Soviet physicist. He was born on 10th February 1894 and died on 23rd January 1952. He is also popular as Jacov Frenkel, well-known for his works on the condensed matter physics. He first proposed the concept of exciton.

Ataç İmamoğlu

Ataç İmamoğlu was born on August 12, 1964. He is a Turkish-Swiss physicist, well-known for his work on quantum optics and quantum computation. In 1993, he joined the University of California, Santa Barbara, as a professor and then in 2001, he moved to the University of Stuttgart, Germany. Since 2002, he has been working as a head of research group of quantum photonics at Swiss Federal Institute of Technology, Switzerland.

First the theory of Polariton's condensation was proposed by a scientist A. Imamoglu [14]. The bosonic nature of polaritons was observed in the experimental work carried out by the group of scientists, Baumberg, Scolnick and Deveaud. Previously it was not possible to see the condensation in the sample because the binding energy of the sample was poor and the system had no strong coupling. The first report of macroscopic occupation in the strong coupling regime came from the samples of R. André which were based on II-VI compounds (CdTe, CdMnTe, CdMgTe). In this sample, exciton's binding energy is high. The research work by scientists Richard *et al.* [15] also reported about condensation but it lacked experimental evidence. In 2006, the first proof of condensation of exciton-polaritons has been reported by the scientists Kasprzak and Richard *et al.* [16]. This work brings new light and excitement to this field. The condensation of exciton-polaritons under stress has been described in GaAs microcavities [17] and recently in planar GaAs microcavities under non-resonant pumping [18]. The evidence of condensation at room temperature has been reported using large bandgap materials like GaN [19] and recently using ZnO materials [20].

SUMMARY OF CHAPTERS

In the first chapter, an introduction to physics of Bose Einstein Condensation of excitons and polaritons with historical sequences of the most important turning points and outcomes has been discussed. The chapters' summaries are given in the next section of chapter 1.

The second chapter is about the fundamentals of excitons. The history of excitons, the name of the scientist who discovered it, procedures of the creation of excitons, types of excitons, properties of excitons etc. have been discussed in the first section of this chapter. From the second section of this chapter, it is possible to know the relaxation kinetics of excitons. In an experiment, generally, excitons are created within a trap with a kinetic energy in the order of meV and then excitons relax down to the center of the trap. In this process, excitons go through different relaxation processes. These are exciton-phonon scattering, exciton-exciton scattering, Auger decay, radiative decay, and non-radiative decay. Generally, the Boltzmann equation has been used to see the relaxation kinetics of excitons theoretically. Here the Boltzman equation with drift and collision interaction terms has been discussed. The equations for different relaxation processes like exciton-phonon scattering, exciton-exciton scattering, Auger decay, radiative decay, and non-radiative decay have been given in detail. The results from the research work about the relaxation kinetics of excitons have been discussed. After going through different relaxation processes, excitons are thermalized within the trap. Thermalization behavior depends on the different conditions especially on the bath temperature of the excitons. In the next section of this chapter, some research results about the thermalization behavior of excitons have been discussed.

Fundamentals of polaritons have been discussed in the third chapter. In the first section of this chapter, the origin of polaritons *i.e.* types of polaritons, procedure of the formation of polaritons, and light matter coupling have been discussed. Special condensed features of polaritons like polariton lasing, super fluidity and quantized vortices have been discussed in the next section.

In the fourth chapter, the Bose Einstein condensation of excitons has been discussed. Recent and early experimental techniques to obtain BEC have been discussed in the first section of this chapter. Spectral analysis experiments and transport experiments are discussed in case of early experiments. Recent experimental procedures have also been discussed in detail. To know the relaxation kinetics and BEC of excitons, the theoretical modeling procedures have been discussed in the next section. Solving the Maxwell Boltzmann equation and Gross Pitaeveski equation are the two ways to obtain the relaxation kinetics and BEC of excitons. Here the Maxwell Boltzmann equation has been discussed in detail and the Gross Pitaeveski equation has been discussed in detail in chapter 5. Then in the next section, the recent developments in this field have been given. In the last section of this chapter, applications of the BEC of excitons have been discussed. Atom laser, atomic clock, and gravitational, rotational, and magnetic sensors are the main applications of the BEC of excitons.

The Bose Einstein Condensation of Polaritons has been discussed in the last chapter, chapter 5. In the first section of this chapter, experimental techniques to get BEC have been discussed. Resonant pumping, non-resonant pumping, and inhomogeneous optical pumping are discussed in detail within the section, excitation methods. General excitation and detection setup also have been discussed. Realspace imaging, momentum space imaging, spectroscopy, and time-resolved imaging have been discussed in detail within the section, optical observation methods. The Gross Pitaeveski equation has been discussed in detail in the next section. In the third section, recent developments in this field have been discussed. Applications of Bose Einstein Condensation of Polaritons have been discussed in the last section of this chapter. Polariton Laser, Polariton switches and quantum simulators are the main applications of the BEC of polaritons.

CONCLUSION

The theoretical idea of BEC was given by Satyendra Nath Bose and Albert Einstein in 1925. 70 years later in 1995, it was first experimentally proven by Eric Cornell and Carl Wieman in a gas of rubidium atoms. After that, BEC of excitons and polariton is a rapidly developing field in solid-state physics. It has possible practical applications in technological devices. For development in the field of quantum information processing, the research on the BEC of excitons and polaritons is essential. The BEC can be used to make quantum computers and it is useful as a quantum simulator.

REFERENCES

[1]   Bose, S.N. Plancks Gesetz und Lichtquantenhypothese. *Eur. Phys. J. A,* **1924**, *26*(1), 178-181.
       http://dx.doi.org/10.1007/BF01327326
[2]   Einstein, A. Quantentheorie des einatomigen idealen Gases. *Preussische Akademieder Wissenshaften,* **1924**, *22*, 261.
[3]   Volovik, G.E.; Mineev, V.P. Line and Point Singularities in Superfluid Helium III. *JETP Lett.,* **1976**, *24*, 561.
[4]   Cross, M.C.; Brinkman, W.F. Textural singularities in the superfluid A phase of3He. *J. Low Temp. Phys.,* **1977**, *27*(5-6), 683-686.
       http://dx.doi.org/10.1007/BF00655703
[5]   Stwalley, W.C.; Nosanow, L.H. Possible "New" quantum systems. *Phys. Rev. Lett.,* **1976**, *36*(15), 910-913.
       http://dx.doi.org/10.1103/PhysRevLett.36.910
[6]   Cornell, E Experiments in dilute atomic bose–einstein condensation *Collected Papers of Carl Wieman,* **2008**, 533-584.
       http://dx.doi.org/10.1142/9789812813787_0075
[7]   Anderson, M.H.; Ensher, J.R.; Matthews, M.R.; Wieman, C.E.; Cornell, E.A. Observation of bose-einstein condensation in a dilute atomic vapor. *Science,* **1995**, *269*(5221), 198-201.
       http://dx.doi.org/10.1126/science.269.5221.198 PMID: 17789847
[8]   Levi, B. Cornell, ketterle, and wieman share nobel prize for bose–einstein condensates. *Search & Discovery, Physics Today,* **2001**.
[9]   Kirtley, J.R.; Tsuei, C.C.; Rupp, M.; Sun, J.Z.; Yu-Jahnes, L.S.; Gupta, A.; Ketchen, M.B.; Moler, K.A.; Bhushan, M. Direct imaging of integer and half-integer Josephson vortices in high-Tc grain boundaries. *Phys. Rev. Lett.,* **1996**, *76*(8), 1336-1339.
       http://dx.doi.org/10.1103/PhysRevLett.76.1336 PMID: 10061695
[10]   Moskalenko, S.A.; Snoke, D.W. *Bose-Einstein Condensation of Excitons and Biexcitons*; Cambridge University Press: Cambridge, **2000**.

http://dx.doi.org/10.1017/CBO9780511721687

[11]   Frenkel, J. On the Transformation of Light into Heat in Solids. II. *Phys. Rev.*, **1931**, *37*(10), 1276-1294.
http://dx.doi.org/10.1103/PhysRev.37.1276

[12]   Frenkel, J. *Soviet Experimental and Theoretical Physics,* **1936**, *6*, 647.

[13]   Weisbuch, C.; Nishioka, M.; Ishikawa, A.; Arakawa, Y. Observation of the coupled exciton-photon mode splitting in a semiconductor quantum microcavity. *Phys. Rev. Lett.*, **1992**, *69*(23), 3314-3317.
http://dx.doi.org/10.1103/PhysRevLett.69.3314 PMID: 10046787

[14]   Imamog⁻lu, A.; Ram, R.J.; Pau, S.; Yamamoto, Y. Nonequilibrium condensates and lasers without inversion: Exciton-polariton lasers. *Phys. Rev. A,* **1996**, *53*(6), 4250-4253.
http://dx.doi.org/10.1103/PhysRevA.53.4250 PMID: 9913395

[15]   Richard, M.; Kasprzak, J.; André, R.; Romestain, R.; Dang, L.S.; Malpuech, G.; Kavokin, A. Experimental evidence for nonequilibrium Bose condensation of exciton polaritons. *Phys. Rev. B Condens. Matter Mater. Phys.*, **2005**, *72*(20), 201301.
http://dx.doi.org/10.1103/PhysRevB.72.201301

[16]   Kasprzak, J.; Richard, M.; Kundermann, S.; Baas, A.; Jeambrun, P.; Keeling, J.M.J.; Marchetti, F.M.; Szymańska, M.H.; André, R.; Staehli, J.L.; Savona, V.; Littlewood, P.B.; Deveaud, B.; Dang, L.S. Bose–Einstein condensation of exciton polaritons. *Nature,* **2006**, *443*(7110), 409-414.
http://dx.doi.org/10.1038/nature05131 PMID: 17006506

[17]   Balili, R.; Hartwell, V.; Snoke, D.; Pfeiffer, L.; West, K. Bose-Einstein condensation of microcavity polaritons in a trap. *Science,* **2007**, *316*(5827), 1007-1010.
http://dx.doi.org/10.1126/science.1140990 PMID: 17510360

[18]   Wertz, E.; Ferrier, L.; Solnyshkov, D.D.; Senellart, P.; Bajoni, D.; Miard, A.; Lemaître, A.; Malpuech, G.; Bloch, J. Spontaneous formation of a polariton condensate in a planar GaAs microcavity. *Appl. Phys. Lett.,* **2009**, *95*(5), 051108.
http://dx.doi.org/10.1063/1.3192408

[19]   Christmann, G.; Butté, R.; Feltin, E.; Carlin, J.F.; Grandjean, N. Room temperature polariton lasing in a GaN/AlGaN multiple quantum well microcavity. *Appl. Phys. Lett.,* **2008**, *93*(5), 051102.
http://dx.doi.org/10.1063/1.2966369

[20]   Sun, L.; Sun, S.; Dong, H.; Xie, W.; Richard, M.; Zhou, L.; Dang, L.S.; Shen, X.; Chen, Z Room temperature one-dimensional polariton condensate in a ZnO microwire. *arXiv10074686,* **2010**.

CHAPTER 2

Fundamentals of Excitons

Abstract: An electron and an electron hole are attracted to each other by electrostatic Coulomb force and combine. This bound state of electron-hole pair is known as an exciton. It can carry energy without transferring net electric charge because it is an electrically neutral quasi-particle. Excitons exist in semiconductors, insulators and some liquids. Frenkel exciton and Wannier-Mott exciton are the two types of excitons. In this section, the origin of excitons, types of excitons, and the relaxation and thermalization behavior of excitons with some research results have been discussed.

Keywords: Auger decay, Exciton phonon scattering, Exciton exciton scattering, Origin of excitons, Relaxation behavior of excitons, Thermalization behavior of excitons.

THE ORIGIN OF EXCITONS

In 1931, Yakov Frenkel first proposed the concept of excitons. He said that without transferring the charge, this excited state can move in a particle-like fashion across the lattice [1, 2].

By using the band theory, it is possible to describe the creation of exciton. At zero temperature, a pure semiconductor has no free charge carriers. Therefore, all the energy levels in the conduction bands are completely vacant and all the energy levels in the valence bands are completely filled with electrons. Excitons can be built in two ways in case of direct band gap semiconductor: the first one is when a semiconductor absorbs a photon that has sufficient energy to give rise to an electron from the valence band to the excitonic bound state but its energy is less than the band gap energy. The second one is when a semiconductor absorbs a photon that has energy equal to the band gap energy or more than this. In this case, an electron from the valence band goes to the conduction band. Then free holes and electrons are created in the valence band and conduction band, respectively. These free holes and electrons bind with each other and produce exciton. Therefore to produce a free electron and hole within a direct band gap semiconductor, the least energy requirement is the band gap energy (Fig. **2.1**).

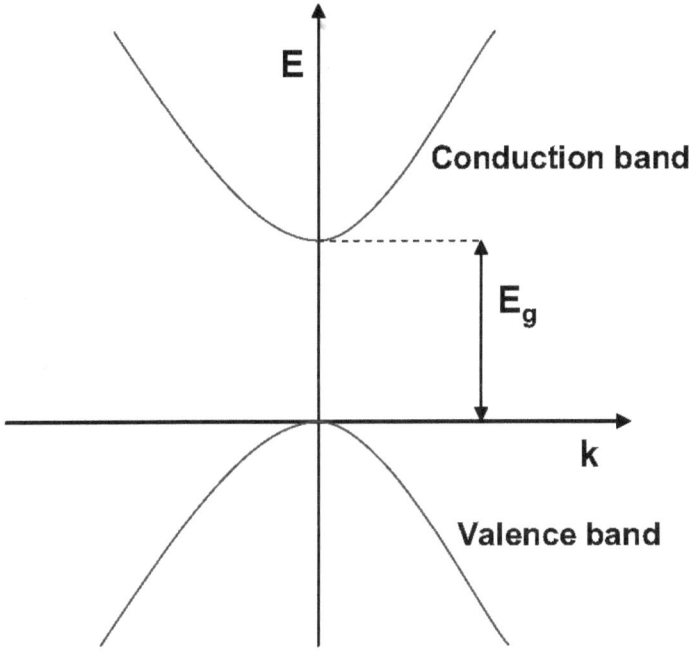

Fig. (2.1). Representation of the band structure of a direct band gap semiconductor. E_g is the band gap energy [3]. E is the energy and k is the momentum vector. Upper side band is known as a conduction band and a lower side band is known as a valence band.

The exciton's binding energy can be represented by the equation:

$$E_b = \frac{e^4 \mu_{ex}}{2\hbar^2 \varepsilon^2 n^2} \qquad n = 1, 2, 3........ \tag{1}$$

Where the exciton reduced mass is μ_{ex}, the electric charge is e, and ε is the static dielectric constant. The relation between the static dielectric constant ε, the dielectric constant $\varepsilon_0 = 8.85419 \times 10^{-12}$ $A^2 s^4 kg^{-1} m^{-3}$ and the dielectric number of the material ε_r is $\varepsilon = 4\pi\varepsilon_0\varepsilon_r$. If the effective mass of the electron is m_e and the effective mass of the hole is m_h then the exciton reduced mass is μ_{ex} that can be written as $\mu_{ex} = m_e m_h / (m_e + m_h)$. The orthoexciton has slightly smaller binding energy than paraexciton due to the spin-orbit coupling. The Hamiltonian of the electron and the hole coupling can be represented by equation [4, 5]:

$$H = -\frac{\hbar^2 \nabla_e^2}{2m_e} - \frac{\hbar^2 \nabla_h^2}{2m_h} - \frac{e^2}{\varepsilon |r_e - r_h|} + \frac{\hbar^2}{2\mu_{ex}} \cdot \frac{l(l+1)}{r^2} \qquad (2)$$

Here the position coordinate of the electron is r_e, the hole is r_h and the exciton is r. The angular momentum quantum number is l. The corresponding Eigen values form a series of exciton energies that can be written as:

$$E_n = E_g - \frac{\mu_{ex}\left(e^2/\varepsilon\right)^2}{2\hbar^2 n^2} + \frac{\hbar^2 k^2}{2M} + \frac{\hbar^2}{2\mu_{ex}} \cdot \frac{l(l+1)}{r^2} \qquad (3)$$

Here the binding energy and the kinetic energy of exciton are represented by the second term and the third term, respectively. The fourth term is also an energy term of exciton that generated during the process of electron hole coupling and the creation of exciton. The band-gap energy is E_g. M is the mass of exciton.

By the emission of photons, excitons make them visible. The excited excitons can come to the ground state by releasing the energy [3]:

$$h\nu = E_g - E_b + E_k \qquad (4)$$

This process is known as the radiative recombination. The radiative recombination is also possible by the emission of phonon and photon together, with energy:

$$h\nu = E_g - E_b + E_k \pm E_p \qquad (5)$$

Here the energy of the phonon is E_p and the exciton kinetic energy is E_k. The "+" sign is used for anti-stokes scattering process and "-" sign is used for stokes scattering process. According to the momentum conservation, two types of processes are possible: one is phonon-assisted transitions $k = k_{photon} + k_{phonon}$ and another is direct transitions $k = k_{photon}$.

The exciton's fine structure is generated by the coupling of spins by the exchange interaction. The hole and electron have either antiparallel or parallel spins. An exciton's property depends on momentum (k-vector) in periodic lattices.

The excitons are of two types, these are Wannier-Mott exciton and Frenkel exciton (Fig. **2.2**). The radius of the exciton is

$$a_E = a_0 \frac{m_e}{\mu_{ex}} \varepsilon$$

$$\text{or, } a_E = \frac{\hbar^2}{\mu_{ex}\left(\dfrac{e^2}{\varepsilon}\right)} \tag{6}$$

Here the electron rest mass is m_e and the Bohr radius is a_0.

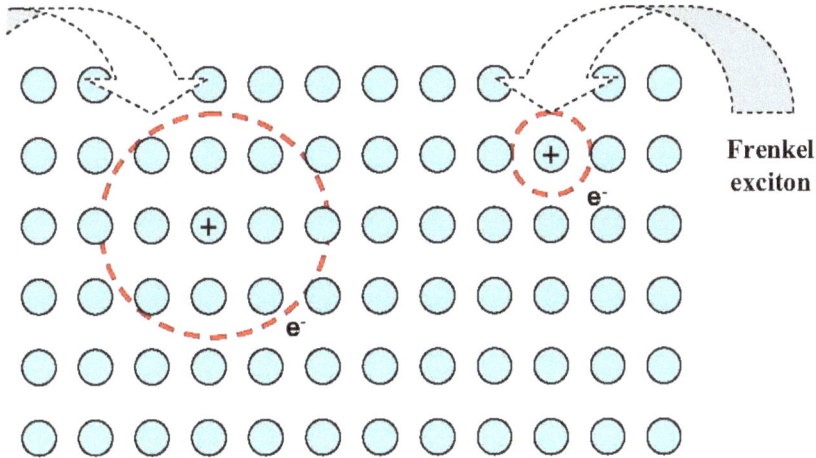

Fig. (2.2). Representation of Wannier-Mott exciton and Frenkel exciton, not to scale [3]. In case of Wannier-Mott exciton, the binding energy between the electron and the hole is approximately 0.01eV and it has a radius larger than the lattice spacing. In case of Frenkel exciton binding energy is in the range of 0.1 to 1 eV and the size is like the unit cell.

Wannier-Mott Exciton: The name Wannier-Mott exciton came from the name of two scientists, they are Gregory Wannier and Nevill Francis Mott. The Wannier-Mott exciton is formed in semiconductors where the electric field reduces the Coulomb interaction and has a large dielectric constant. In this case, the electron and hole attraction is very little in comparison with the band-gap energy. These excitons are weakly bound and are called Wannier-Mott exciton. It has a radius larger than the lattice spacing. Because of its low Coulomb interaction and the lower masses, its binding energy is approximately 0.01eV that is much less than that of a hydrogen atom. The Wannier-Mott exciton also known as large excitons, is found in semiconductor crystals such as cuprous oxide (Cu_2O), zinc selenide (ZnSe), gallium arsenide (GaAs), copper chloride (CuCl) *etc*. It is also found in some liquids, such as liquid Xenon.

Frenkel Exciton: The name Frenkel exciton, came from the scientist Yakov Frenkel. In case of strong Coulomb interaction between an electron and a hole and relatively small dielectric constant, Frenkel exciton is formed in some materials. In this case, exciton is strongly bound and the radius is on the order of the lattice constant. Its binding energy is in the range of 0.1 to 1 eV and the size is like the unit cell. Some inorganic molecular crystals contain aromatic molecules like anthracene and tetracene. The Frenkel excitons are formed within these inorganic molecular crystals and alkali halide crystals. In transition metals containing compounds with partially-filled *d*-shells, the on-site *d-d* excitation is another example of Frenkel exciton. Generally due to symmetry, the *d-d* transitions are forbidden. But by structural relaxations or other effects when the symmetry is broken, the *d-d* transitions are weakly-allowed. An electron-hole pair is formed when a *d-d* transition absorbs photons. This electron-hole pair can be treated as a Frenkel exciton.

RELAXATION KINETICS OF EXCITONS

Generally in an experimental work, excitons are produced within a potential trap with the kinetic energy of the range of meV. After creation, excitons undergo different relaxation processes and come to the center of the potential trap. To numerically calculate the relaxation kinetics of excitons, the Boltzmann equation can be used.

The Boltzmann equation is represented by the equation

$$\frac{\partial S}{\partial t} + \vec{v} \cdot \overrightarrow{\nabla_r} S + \vec{F} \cdot \frac{1}{\hbar} \overrightarrow{\nabla_k} S = \left(\frac{\partial S}{\partial t} \right)_{\text{collision+interaction}} \tag{7}$$

The number of particles as a function of radius, time and momentum can be explained by the Boltzmann equation. In equation (7), $S(\vec{r}, \vec{k}, t)$ is the number of particle, $\vec{r}$ is the radius, $\hbar\vec{k}$ is the momentum and t is the time. The force is represented by $\vec{F}$, the velocity is denoted as $\vec{v}$ and the nabla operators are ∇_r and ∇_k. The collision and interaction terms are on the right hand side of the equation (7). The exciton-exciton scattering, Auger decay, exciton-Phonon scattering, radiative and non-radiative decay are the collision and interaction terms. The drift terms are the left part of the equation. To do the numerical calculations and to see the relaxation dynamics, first excitons are produced which after going through the different relaxation process, come back to the bottom of the trap.

Exciton-Phonon Scattering

In 1950, first the concept of the exciton-phonon collision emerged. The exciton phonon collision is the reaction between a nonpolar optical phonon and an electron. Due to this collision, a small displacement in the electronic energy band of the crystal takes place that is known as the lattice displacement of the electron. The deformation potential is the proportionality constant between the lattice displacement and the energy shift. In a direct band gap semiconductor, the interaction of excitons with acoustic phonons gives the temperature-dependent part of the exciton distribution. In an ionic crystal for the interaction between excitons and lattice vibration, two types of mechanisms mainly work. One is the polarization or Fröhlich interaction and another is the short-range deformation interaction. Due to the Coulomb interaction between the charge carriers, Fröhlich interaction takes place. The deformation interaction occurs due to the reformation of the excitons wave function by longitudinal vibrations. The deformation potential is the magnitude of the interaction. The dynamics of excitons and the energy spectrum of excitons for relatively small radius are strongly affected by the deformation interaction.

The dynamics of a homogeneous system of excitons can be described by the exciton-Phonon scattering term. The exciton-phonon collision term of the Boltzmann equation [6] is represented by equation [3].

$$\frac{\partial S_{\vec{k}}}{\partial t} = -\frac{2\pi}{\hbar} \sum_{\vec{p}} \left| M_{x-ph}(\vec{p}-\vec{k}) \right|^2 \{ [S_{\vec{k}}(1+n_{\vec{k}-\vec{p}}^{ph})(1+S_{\vec{p}}) - (1+S_{\vec{k}})n_{\vec{k}-\vec{p}}^{ph} S_{\vec{p}}] \delta(\varepsilon_{\vec{k}} - \varepsilon_{\vec{p}} - \hbar\omega_{\vec{k}-\vec{p}}) \tag{8}$$
$$+ [S_{\vec{k}} n_{\vec{p}-\vec{k}}^{ph}(1+S_{\vec{p}}) - (1+S_{\vec{k}})(1+n_{\vec{p}-\vec{k}}^{ph})S_{\vec{p}}] \times \delta(\varepsilon_{\vec{k}} - \varepsilon_{\vec{p}} + \hbar\omega_{\vec{p}-\vec{k}}) \} - S_{\vec{k}} / \tau_{opt}$$

Here $\varepsilon_{\vec{k}} = \hbar^2 k^2 / 2M_x$ stands for the exciton energy in $\vec{k}$ state, $\varepsilon_{\vec{p}} = \hbar^2 p^2 / 2M_x$ stands for the exciton energy in $\vec{p}$ state and $\hbar\omega_{\vec{p}-\vec{k}} = \hbar v_s |\vec{p}-\vec{k}|$ is the phonon energy. The exciton occupation number in $\vec{k}$ state is symbolised by $S_{\vec{k}}$, exciton occupation number in $\vec{p}$ state is symbolised by $S_{\vec{p}}$ and the phonon occupation number is $n_{\vec{p}-\vec{k}}^{ph} = 1 / [\exp(\hbar\omega_{\vec{p}-\vec{k}} / k_B T_b) - 1]$. $M_{x-ph}(\vec{p}-\vec{k})$ stands for the matrix element of the exciton-phonon deformation potential interaction and the exciton's radiative lifetime is τ_{opt}. The term $\left|M_{x\ ph}(\vec{p}-\vec{k})\right|^2 = \hbar D^2 \left|\vec{p}-\vec{k}\right|/(2V\rho v_s)$ is known as the exciton-phonon coupling. Where the exciton mass is M_x, the deformation potential energy is denoted by D, the longitudinal acoustic sound velocity is v_s, the crystal volume and density are V and ρ respectively. The Dirac delta distribution or delta function is δ. The Stokes scattering is given by $[S_k(1+n_{k-\vec{p}}^{ph})(1+S_{\vec{p}}) - (1+S_k)n_{k-\vec{p}}^{ph}S_{\vec{p}}]$ and the anti-Stokes scattering of excitons is given by $[S_{\vec{k}}n_{\vec{p}-\vec{k}}^{ph}(1+S_{\vec{p}}) - (1+S_{\vec{k}})(1+n_{\vec{p}-\vec{k}}^{ph})S_{\vec{p}}]$.

Exciton-Exciton Scattering

Exciton-exciton scattering, also known as elastic scattering, is a collision and interaction term of the Boltzmann equation [7]. The elastic collision is the two-body collision of the boson gas. In case of exciton phonon collision, one initial exciton momentum and one final exciton momentum take part in but two initial exciton momenta and two final exciton momenta take part in case of elastic collision. In this case, excitons of momentum p2 and p collide with momentum k2 and k and the collision rate takes account of all events. The rate of scattering of the exciton into state k is [3, 8, 9], which is given by:

$$\left(\frac{dS_k}{dt}\right)_{elastic\_in\_scattering} d^3\vec{k} = \frac{2\pi}{\hbar} \cdot \frac{V^2}{(2\pi)^6} \int d^3\vec{k}\, d^3\vec{p}\, d^3\vec{p}_2\, d^3\vec{k}_2\, M_{matrix}^2$$

$$\times \{S_{\vec{p}}S_{\vec{p}_2}(1+S_{\vec{k}})(1+S_{\vec{k}_2})\}\delta(\vec{p}+\vec{p}_2-\vec{k}-\vec{k}_2)$$

$$\times \delta(\varepsilon_{\vec{p}}+\varepsilon_{\vec{p}_2}-\varepsilon_{\vec{k}}-\varepsilon_{\vec{k}_2}) \tag{9}$$

Where the exciton energy, in wave vector $\vec{k}$ state is given by $\varepsilon_{\vec{k}} = \hbar^2 k^2/2M_x$, in wavevector $\vec{p}$ state, it is symbolised by $\varepsilon_{\vec{p}} = \hbar^2 p^2/2M_x$, in wavevector $\vec{p}_2$ state, it is symbolised by $\varepsilon_{\vec{p}_2} = \hbar^2 p_2^2/2M_x$ and in wavevector $\vec{k}_2$ state, it is given by $\varepsilon_{\vec{k}_2} = \hbar^2 k_2^2/2M_x$. The exciton occupation number in $\vec{k}$ state, $\vec{p}$ state, $\vec{p}_2$ state, and $\vec{k}_2$ state is $S_{\vec{k}}$, $S_{\vec{p}}$, $S_{\vec{p}_2}$ and $S_{\vec{k}_2}$ respectively. The matrix element is represented by $M_{matrix} = \dfrac{4\pi\hbar^2 a_1}{M_x V}$ [9] where the scattering length is a_1. Here $\varepsilon_{k_2} = \varepsilon_{\vec{p}} + \varepsilon_{\vec{p}_2} - \varepsilon_k$

or $\vec{k}_2 = \sqrt{\vec{p}^2 + \vec{p}_2^2 - \vec{k}^2}$.

Auger Decay

Auger recombination is a three body process. Where first the electrons and holes remain unbound, and then one electron-hole couple reunites and donates the band gap energy to the excess electron or hole. In Cu₂O and at low temperatures, by reunion of one exciton and ionization of the other one, two excitons are demolished by Auger decay [10]. For calculation, one can assume that the new excitons are formed from the rebinding of the ionized carriers produced by Auger decay. These excitons are allocated over the whole energy range. The orthoexcitons are created from the recombination of the electron hole pairs then within 3 ns, these are converted into the paraexcitons. Therefore, half of the excitons are recovered that were demolished by Auger decay. The change of the exciton occupation number due to Auger decay is [10] given by:

$$\left(\frac{\partial S_k}{\partial t}\right)_{\text{Auger Decay}} = -A_{pp} f(\vec{r}) S(\vec{r},\vec{k}) + \frac{1}{2} A_{pp} f(\vec{r})^2 \cdot \frac{(2\pi)^3}{\int d^3\vec{k}} \sqrt{a^2 + b^2} \qquad (10)$$

or,

$$\left(\frac{\partial S_e}{\partial t}\right)_{\text{Auger Decay}} = -A_{pp} f(\vec{r}) S(\vec{r},e) + \frac{1}{2} A_{pp} f(\vec{r})^2 \cdot \frac{(2\pi)^3}{\displaystyle\int_0^{e_{max}} \frac{4\sqrt{2}\pi M_x^{3/2} \sqrt{e}}{\hbar^3} de}$$

$$= -A_{pp} f(\vec{r}) S(\vec{r},e) + \frac{1}{2} A_{pp} f(\vec{r})^2 \cdot \frac{3\hbar^3 (2\pi)^3}{8\sqrt{2}\pi M_x^{3/2} e_{max}^{3/2}} \qquad (11)$$

The Auger constant is A_{pp} and the local density is $f(\vec{r})$. The first term stands for the two body decay and the recovered excitons are represented by the second term. Due to two body decay, the decompose rate of excitons is An^2 where A stands for the Auger constant and n is the gas density.

By taking the total number of exciton along the energy path, the local density can be calculated by:

$$f(\vec{r}) = \frac{1}{(2\pi)^3} \int_k 4\pi k^2 S(\vec{r},\vec{k})dk \tag{12}$$

or,

$$f(\vec{r}) = \frac{1}{(2\pi)^3} \int_e \frac{4\sqrt{2}\pi M_x^{3/2}\sqrt{e}}{\hbar^3} \cdot S(\vec{r},e)de \tag{13}$$

One experimental work [11] used an Auger constant $A_{pp} = 10^{-18}$ cm^3 / ns .

Radiative and Non-radiative Decay

The radiative and non-radiative decay [11] of excitons can be calculated by the equation below:

$$\left(\frac{\partial S_e}{\partial t}\right)_{R\_NR\ decay} = -\Gamma_p S(\vec{r},e) \tag{14}$$

Here the total decay rate is Γ_p. From one experimental work, the decay rate is $\Gamma_p = 1/650$ ns^{-1} [11].

Results and Discussion with Different Relaxation Processes

In several research works, it has been seen that after excitations, excitons cool down and gather in a position close to the bottom of the trap (Fig. **2.3**) [12, 13]. Fig. **(2.3)** represents the numerical simulation results with exciton-phonon interaction for different times at 0.5 K. The series of figures show that how excitons are gathered at the bottom of the trap within 100 ns at 0.5 K. The authors of reference [13] studied the relaxation kinetics of excitons. For the volume of 1 μm^3, the exciton occupation n_r vs. energy for different times at 0.1 K has been shown in Fig. **(2.4)**. The exciton occupation number n_r has been calculated by taking the total exciton number along the radius path. In this case, the exciton-phonon and the exciton-exciton scattering have been included. Here it has been shown that high exciton occupation number is near zero energy. This is one important criterion to obtain

BEC. But if Auger decay and radiative non radiative decay are included with exciton-phonon scattering, then no high exciton occupation is observed near zero energy (Fig. **2.5**).

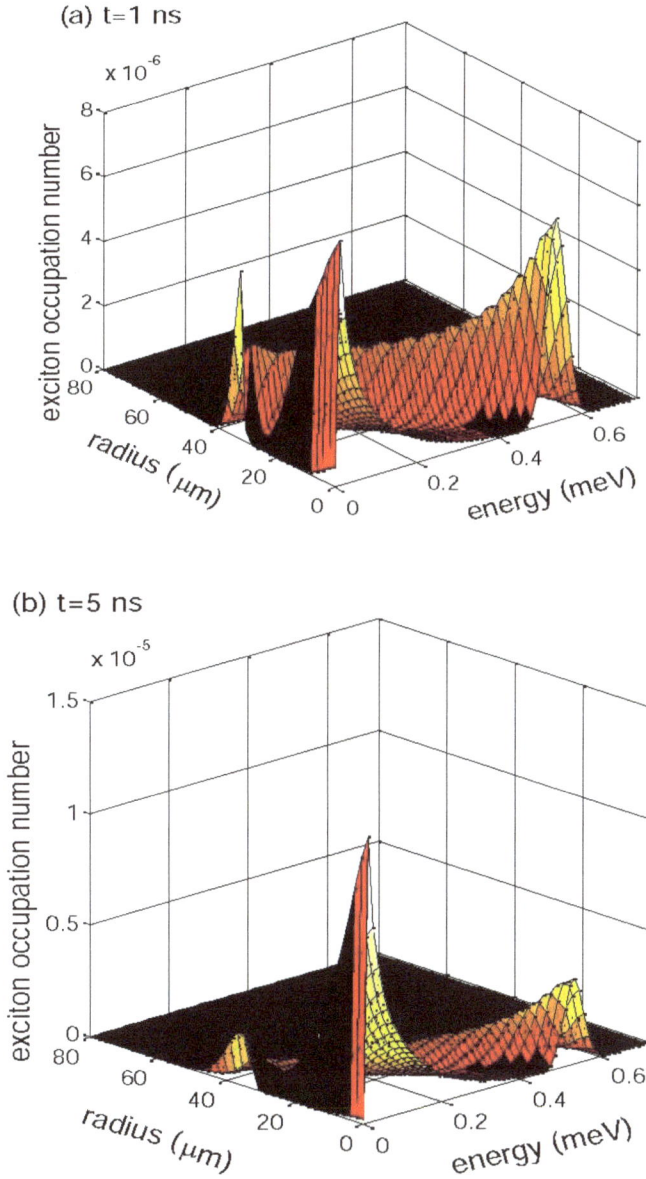

(a) t=1 ns

(b) t=5 ns

(Fig. 2.3) contd.....

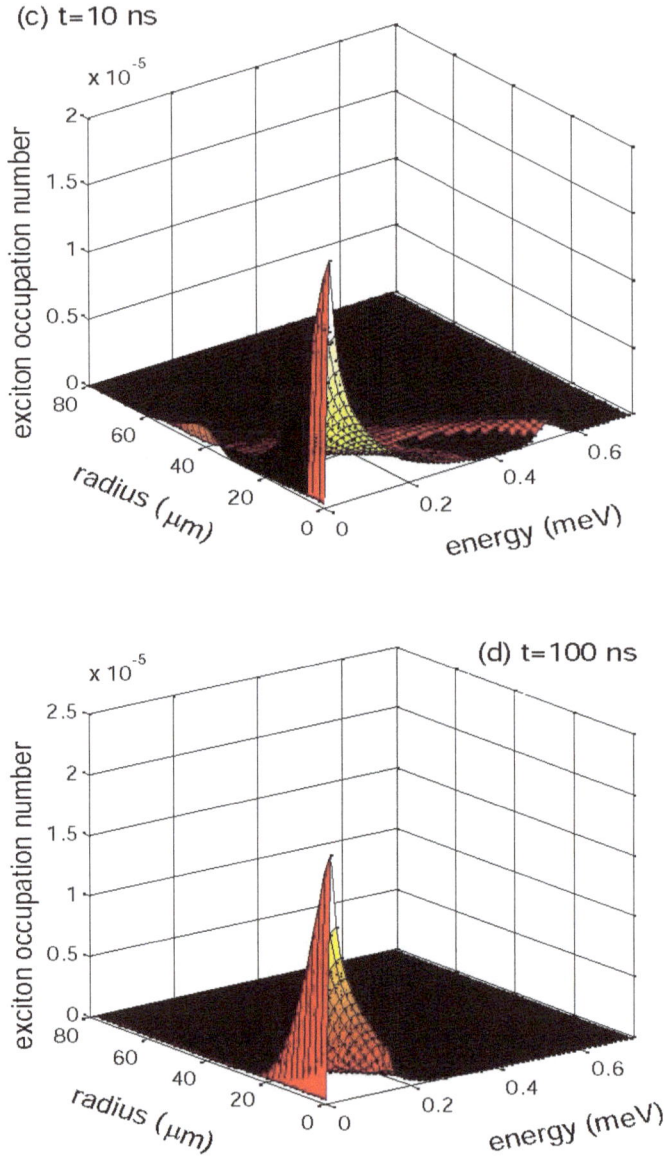

Fig. (2.3). The numerical simulation results with exciton-phonon interaction for different times at 0.5 K. In this case, 1.7×10^4 initial exciton number has been taken. The value of the parameter is $M_x = 2.61\, m_e$, $m_e = 9.109 \times 10^{-31}\, kg$ is the electron rest mass, $V_s = 4.5 \times 10^3\, \text{m/s}$, $a = 0.5 \cdot \mu eV \cdot \mu m^2$, D=1.68 eV and $\rho = 6.11 \times 10^3\, kg/m^3$. The series of figures indicates that excitons are gathered at the bottom of the trap within 100 ns at 0.5 K. Reprinted (adapted) with permission from [13].

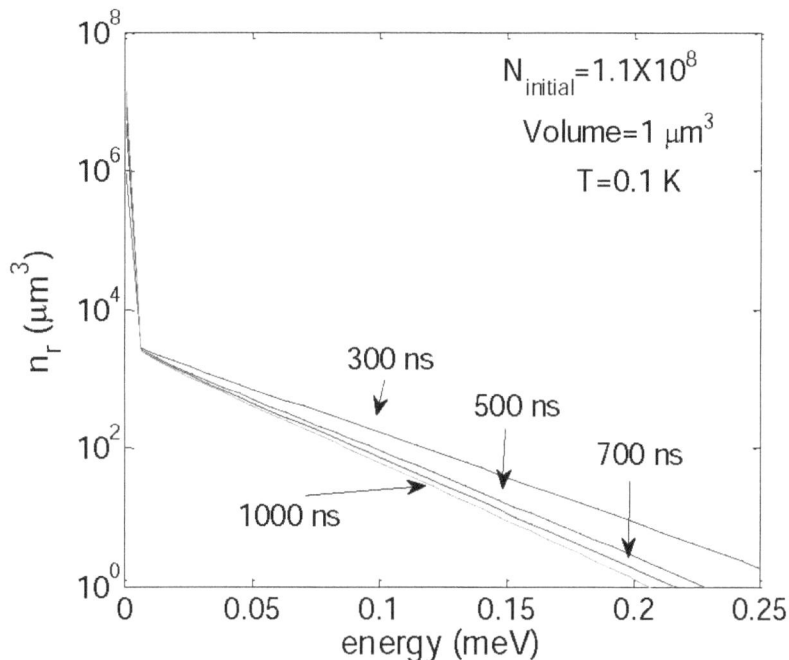

Fig. (2.4). The total number of excitons (nr) along the radius *vs.* energy for different times at 0.1 K for the volume 1 μm³. In this case 1.1×10^8 is the initial exciton number [14] and as relaxation processes, the exciton-exciton and exciton-phonon scattering have been taken. In these cases, a high exciton occupation number is observed near zero energy for each time that is one important condition to obtain BEC.

THERMALIZATION BEHAVIOUR OF EXCITONS

Thermalization behaviour of excitons depends on different conditions like temperature, relaxation processes. The authors of reference [13] studied the thermalization behaviour of excitons. For non-degenerate cases, they have computed the local effective temperature T_{local} to investigate the thermalization behaviour of excitons. The long energy tail of the exciton allocation is fitted with the function $e^{-e/k_B T_{local}}$. They calculated T_{local} and studied its changes with time. To get the exciton distribution, the total number of excitons (n_r) along the radius has been calculated. Then $\log(n_r)$ *vs.* energy curves have been plotted and fitted with the function $e^{-e/k_B T_{local}}$. After that, T_{local} was calculated for different bath temperatures (Fig. **2.6**).

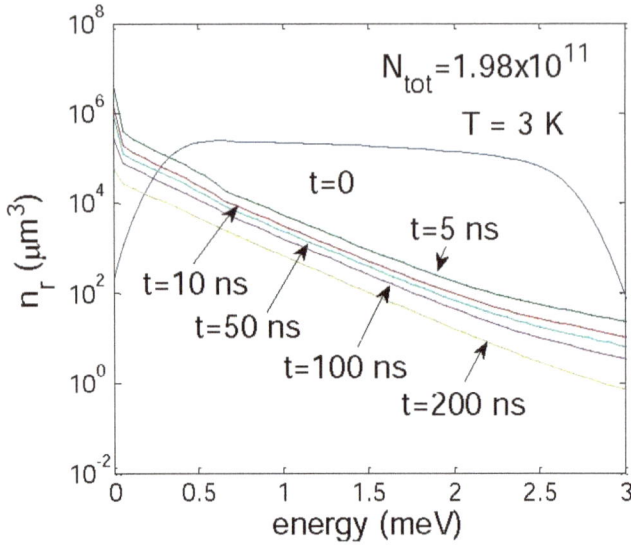

Fig. (2.5). The total exciton number of non-degenerate exciton gas (nr) along the radius *vs.* energy at different times at 3 K. Initially 1.98×10^{11} is the total exciton number [14]. In this case, the Auger decay and radiative non-radiative decay are included with the exciton-phonon scattering. In this case, no high exciton occupation is observed near zero energy.

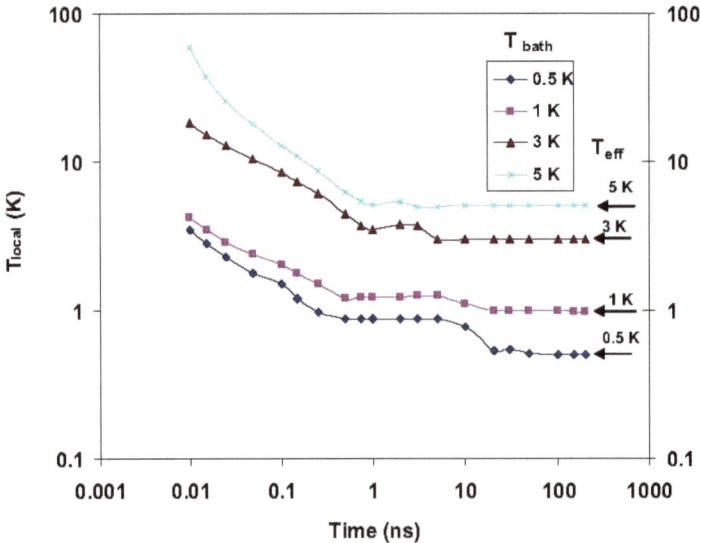

Fig. (2.6). The local temperature of Exciton T_{local} *vs.* time for the different bath temperatures including exciton-phonon scattering. It has been observed that the effective temperature changes with changing time for the different bath temperatures. The arrows on the right hand side indicate effective temperatures. Initially, the total number of excitons is 1.5×10^4, 1.5×10^4, 1.98×10^6 and 1.99×10^6 at temperatures of 0.5 K, 1 K, 3 K and 5 K respectively. Here it has been observed that at higher temperatures like 3 K and 5K, the effective temperature reduces to the bath temperature within 10 ns but for temperature below 1 K, it takes around hundreds of nanoseconds. Reprinted (adapted) with permission from [13].

From the figure it can be noted that the effective temperature reaches the bath temperature very slowly for the temperature below 1K. It takes around hundreds of nanoseconds. The effective temperature reaches the bath temperature within ten nanoseconds for the temperature above 1K.

From the results of reference [13], it has been observed that at temperature 3 K, 1.5 K and 0.5 K, if the Auger decay and radiative non radiative decay are included with the exciton phonon scattering, then the effective temperatures are not able to reach the bath temperatures even after 600 ns (Fig. **2.7**). In these cases, the effective temperatures could not be able to reach bath temperatures because of the heat generated during the Auger decay process.

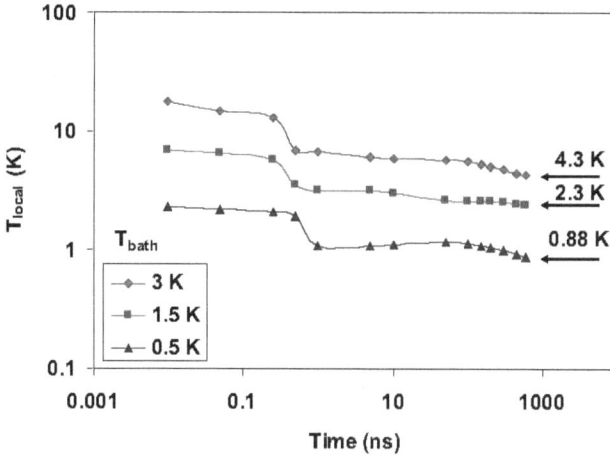

Fig. (2.7). The local effective temperature of Exciton T_{local} *vs.* time curves for different bath temperatures. In this case, Auger decay, radiative decay, non-radiative decay and exciton- phonon scattering have been added as relaxation processes. The arrows indicate effective temperatures. It has been observed that even after 600 ns, the effective temperatures are not able to reach bath temperatures. The initial exciton number within the trap is 1.02×10^9 at 0.5 K, 2.76×10^{10} at 1.5 K and 1.99×10^{11} at 3 K. Reprinted (adapted) with permission from [13].

In a study [15], it has been observed that if the exciton-exciton scattering has been added to the exciton-phonon scattering but no Auger decay and radiative and non-radiative decay, then the effective temperatures reach the bath temperatures for 3 K and 0.5 K but it stays little higher than the bath temperature for 0.1 K (Fig. **2.8**). This may be due to the freezing of phonons at low temperatures.

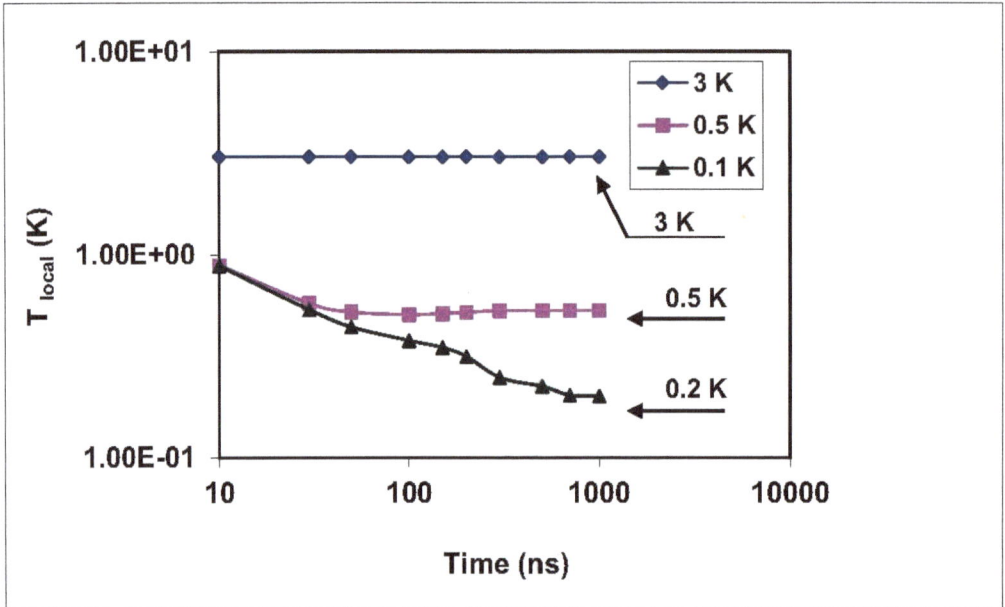

Fig. (2.8). The local temperature of Exciton T_{local} *vs.* time for different bath temperatures. In this case, exciton-exciton and exciton-phonon scattering have been included. Here it has been observed that for the different bath temperatures, with changing time, the effective temperature also changes. The arrows on the right hand side indicate effective temperatures. Initially, the total number of exciton is 1.16×10^8 for 0.5 K and 3 K and 1.1×10^8 for 0.1 K. For 0.1 K, the effective temperature is not reduced to bath temperature even after 1000 ns but for 0.5 K and 3 K, the effective temperatures reduce to bath temperatures. Reprinted (adapted) with permission from [15].

CONCLUSION

Boltzmann equation can be used to see the relaxation and thermalization behavior of excitons. Within a trap after excitations, the excitons experience different relaxation processes like exciton-phonon scattering, exciton-exciton scattering, auger decay and radiative decay. Then excitons come to the bottom of the trap. It has been seen that with exciton phonon scattering, the effective temperatures of excitons reach the bath temperatures within 10 ns for the temperatures over 1 K but in the case of temperature below 1 K, it takes around 100 ns. If Auger decay and radiative decay is considered with exciton-phonon scattering, then the local effective temperatures of excitons do not reach the bath temperatures at any temperature due to heating generated in the Auger process. When exciton-exciton scattering is considered with exciton-phonon scattering, then the effective temperatures of excitons reach bath temperatures at 0.5 K and 3 K but for 0.1 K, it remains little higher than the bath temperature. This may be due to the freezing out of phonons at ultra-low temperatures.

REFERENCES

[1]   Frenkel, J. On the transformation of light into heat in solids. II. *Phys. Rev.,* **1931,** *37*(10), 1276-1294.
      http://dx.doi.org/10.1103/PhysRev.37.1276

[2]   Frenkel, J. *Soviet experimental and theoretical physics,* **1936,** *6,* 647.

[3]   Som, S. *Numerical simulation of exciton dynamics in cuprous oxide at ultra low temperatures.* Ph. D. thesis, University of Rostock: Germany, **2015.**

[4]   Karpinska, K.; van Loosdrecht, P.H.M.; Handayani, I.P.; Revcolevschi, A. Para-excitons in —a new approach. *J. Lumin.,* **2005,** *112*(1-4), 17-20.
      http://dx.doi.org/10.1016/j.jlumin.2004.09.038

[5]   Libo, R.L. *Introductory quantum mechanics*; Addison Wesley: San Francisco, **2003.**

[6]   Ell, C.; Ivanov, A.L.; Haug, H. Relaxation kinetics of a low-density exciton gas in Cu_2O. *Phys. Rev. B Condens. Matter,* **1998,** *57*(16), 9663-9673.
      http://dx.doi.org/10.1103/PhysRevB.57.9663

[7]   Kadanoff, L.P.; Baym, G. *Quantum Statistical Mechanics*; Addison-Wesley: New York, **1989.**

[8]   Snoke, D.W.; Wolfe, J.P. Population dynamics of a Bose gas near saturation. *Phys. Rev. B Condens. Matter,* **1989,** *39*(7), 4030-4037.
      http://dx.doi.org/10.1103/PhysRevB.39.4030 PMID: 9948737

[9]   Snoke, D.W.; Rühle, W.W.; Lu, Y.C.; Bauser, E. Evolution of a nonthermal electron energy distribution in GaAs. *Phys. Rev. B Condens. Matter,* **1992,** *45*(19), 10979-10989.
      http://dx.doi.org/10.1103/PhysRevB.45.10979 PMID: 10001019

[10] O'Hara, K.E.; Wolfe, J.P. Relaxation kinetics of excitons in cuprous oxide. *Phys. Rev. B Condens. Matter,* **2000,** *62*(19), 12909-12922.
      http://dx.doi.org/10.1103/PhysRevB.62.12909

[11] Schwartz, R.; Naka, N.; Kieseling, F.; Stolz, H. Dynamics of excitons in a potential trap at ultra-low temperatures: Paraexcitons in Cu_2O. *New J. Phys.,* **2012,** *14*(2), 023054.
      http://dx.doi.org/10.1088/1367-2630/14/2/023054

[12] Snoke, D.; Kavoulakis, G. M. Bose–Einstein condensation of excitons in Cu_2O: Progress over 30 years *Rep rog Phys,* **2014,** *77,* 116501.
      http://dx.doi.org/10.1088/0034-4885/77/11/116501

[13] Som, S.; Kieseling, F.; Stolz, H. Numerical simulation of exciton dynamics in Cu_2O at ultra-low temperatures within a potential trap. *J. Phys. Condens. Matter,* **2012,** *24*(33), 335803.
      http://dx.doi.org/10.1088/0953-8984/24/33/335803 PMID: 22836306

[14]   Som, S. *Numerical simulation of exciton dynamics in cuprous oxide at ultra low temperatures*. PhD Thesis, The University of Rostock: Germany, **2015**.

[15]   Som, S. Relaxation and condensation kinetics of trapped excitons at ultra-low temperatures: Numerical simulation. *Indian J. Phys. Proc. Indian Assoc. Cultiv. Sci.*, **2020**, *94*(10), 1603-1613.

http://dx.doi.org/10.1007/s12648-019-01592-7

CHAPTER 3

Fundamentals of Polaritons

Abstract: Polariton is a bosonic quasiparticle. When an electromagnetic wave strongly interacts with the dipole active transition of the medium, then polariton is formed [1]. In other words, when light scatters and goes through any interrelated resonance, then the polaritons can be formed. In this chapter, the history of polariton, category of polaritons, their constituents, and light-matter interaction have been explained. Special condensate behaviour of polaritons like polariton lasing, superfluidity, and quantized vortices are also explained.

Keywords: Building block of polaritons, History of polaritons, Light-matter coupling, Polariton laser, Superfluidity, Quantized vortices.

THE ORIGIN OF POLARITONS

The coupling between a photon and a polar excitation in a substance is known as polariton (Fig. **3.1 A**). Polaritons are of different types. They are phonon polaritons, exciton, Intersubband polaritons, Surface Plasmon polaritons, Bragg polaritons, Plexcitons, Magnon polaritons, pi-tons, and Cavity polaritons.

- **Phonon polariton:** When the coupling of an infrared photon and an optical phonon takes place, then the Phonon polariton is formed.
- **Exciton polariton:** When the interaction of exciton with visible light takes place, then the exciton polariton is formed [2].
- **Intersubband polariton:** The intersubband polariton is formed due to the coupling of an intersubband excitation with an infrared or terahertz photon.
- **Surface Plasmon polaritons:** When the Surface Plasmons interact with light, then the surface Plasmon polaritons are formed.
- **Bragg polaritons:** When the coupling of bulk excitons and Bragg photon modes takes place, then the Bragg polariton is formed. It is also known as Braggoritons [3].
- **Plexcitons:** The Plexcitons are formed due to the coupling of plasmons and excitons [4].
- **Magnon polaritons:** When magnons interact with light, then the magnon polaritons are formed.
- **Pi-tons:** When interchanging charge or spin variations interact with light, then the pi-tons are formed. This is completely dissimilar from magnon or exciton polariton.

- **Cavity polaritons:** Within a semiconductor microcavity, both the exciton and photon have a 2D character. When these excitons and photons interact, the cavity polaritons are formed.

Scientist Tonks and Langmuir first observed the oscillation in ionized gasses in 1929 [5]. Then in the early 1950, the classical theory of electromagnetic waves in the ionic crystal near optical phonon frequency was proposed by Tolpygo [6, 7] in 1950 in a Soviet scientific literature. First, they called it a light exciton which was recommended by the Scientist Pekar. The name polariton was first suggested by the Scientist Hopfield which was accepted later. The combined states of electromagnetic waves and phonons in ionic crystals are called the phonon polariton. Dispersion relation of the phonon polariton was obtained by Tolpygo in 1950 [6, 7] and Huang in 1951 [8, 9]. Later on, a study on quantum theory of electromagnetic waves near excitonic resonance and polariton was conducted by Scientists Fano in 1956, Hopfield in 1958 and Agranovich in 1959. In 1952, Scientists Pines and Bohm published their work on collective interaction. In 1955, Scientists Frohlich and Pelzer described Plasmons within silver. Then in 1962, Scientists Ritchie and Eldridge published their experimental work on released photons from the irradiated metal foils. In 1968, the Scientist Otto published a paper on surface Plasmon-polaritons [10]. In 2016, the superfluidity of polaritons at room temperature was noticed by using an organic microcavity with stable Frenkel exciton-polaritons by Giovanni Lerario *et al.* [11]. The invention of a new three-photon bound state has been reported by some scientists[12, 13] in 2018. In the advancement of quantum computers, these new three-photons related to polaritons could be helpful.

Building Blocks for Polariton Formation

Exciton in Semiconductors

When an electron couples with a hole by coulomb interaction then a bosonic quasiparticle exciton is formed. An excited electron comes to the conduction band from the valence band and produces a hole. Then, this electron and hole bind together and make the exciton. Electron's energy can be described as:

$$E_e(k) = E_g + \frac{\hbar^2 k^2}{2m_{e\ eff}}$$

(1)

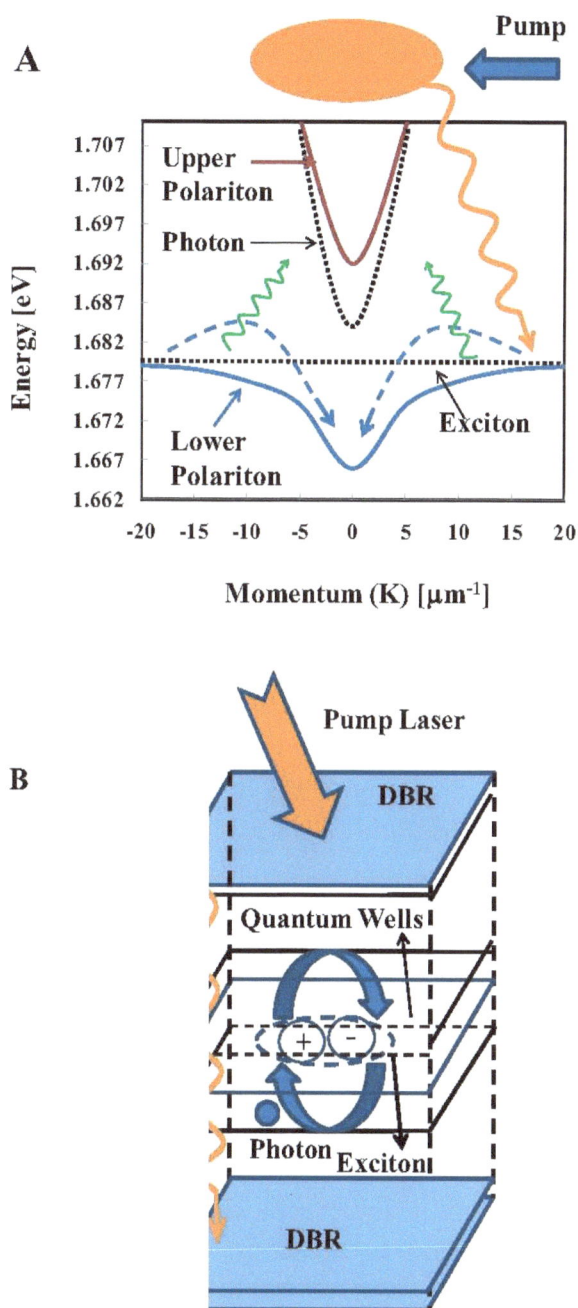

Fig. (3.1). (A) Upper polariton branch, lower polariton branch and polaritons optical excitation scheme represented in the energy and momentum vector (k) plane. Here, hot electron-hole pairs are injected into the microcavity by laser pumping. Excitons are produced and strongly interact with the cavity photons. Then polaritons (upper polariton branch and lower polariton branch) are formed at the time of entering in the light cone. When excitons cool down, excess energy emits through the exciton-phonon and exciton-exciton scatterings. (B) Microcavity polariton. A microcavity is a planar Fabry–Perot resonator including two distributed Bragg Reflectors (DBR) at resonance with excitons in quantum wells. Within microcavities, in the strong coupling region of the light–matter interaction, excitons and cavity photons interact and produce microcavity polaritons.

Where E_g is the energy band gap of the lattice material, m_{e_eff} is the effective mass of the excited electron in the crystal lattice, $\hbar$ is the planck's constant and k is the wave vector of the particle.

When an excited electron comes to the conduction band from the valence band, it leaves a hole in the valence band. The hole has the same momentum as an excited electron but with the opposite sign. The energy of the hole is given as:

$$E_h(k) = E_g + \frac{\hbar^2 k^2}{2m_{h\_eff}}$$

(2)

A free exciton is formed when the particle's momentum is very low. Free excitons can be considered as an equivalent to the hydrogen atom when it transmits in the host lattice of the solid. The details about excitons are provided in chapter 2.

Quantum Well Exciton

A semiconductor quantum well is a thin layer of semiconductor with a thickness that is close to the length of the exciton Bohr radius. It represents a finite quantum mechanical potential well for electronic particles in solids. For motion in space, it also reduces their degrees of freedom into 2D from 3D in case of quantum wells, in bulk matter. In this case, excitons will be confined to one direction [14-18]. In-plane propagation of excitons is allowed if the potential is made by a thin layer of semiconductor material inserted between the barrier layers while the electrons and holes are both not allowed to enter into the surrounding barrier layers. These barrier layers are generated in a higher band gap material. Therefore if the confinement direction is Z, then it will be trapped less in all perpendicular directions of the quantum well and can be estimated as a square well potential along the Z direction. Then only the least energy quantized level of the center-of-mass motion will be useful when the confinement is strong enough [17]. In this case, quantum wells act as two-dimensional quasiparticles. Momentum conservation in case of an optical transition should be fulfilled only in the quantum well plane but not in the Z direction. They can interplay with light with the same in-plane wave number k_{xy}

and arbitrary transverse wave number k_z. A quantum well exciton has a small Bohr radius around half of the size of a bulk exciton [19] as it is within a confinement zone, and it has four times larger binding energy [20]. Therefore a quantum well exciton is stronger than a bulk exciton and the probability of excitation of an absorbed photon to an exciton is also increased with respect to the unconfined case. Optical activity is increased due to this behaviour of quantum well excitons that inspire implant quantum wells in microcavities to increase the light-matter coupling [21].

The width of the thin potential is on the order of the exciton Bohr radius. In the quantum well materials, it has to be chosen in such a way that the thickness will match with the de-Broglie wavelength of electrons and holes. The exciton loses its bulk and isotropic properties in 3D because its properties firmly depend on the width of the confinement potential. The electron and hole wave functions overlap more and more with the generated excitons when the thickness squeezes into the quantum well plane. Oscillator strength and binding energy also increase with these excitons. One important feature of quantum wells is that charge carriers can not automatically overcome the potential barrier without transferring energy to the confined particle. Charge carriers can overcome the potential barrier *via* tunneling effect. This is also an important property of quantum wires that represent 1 D confinement potential and quantum dots which represent 0 D confinement potential. Otherwise, the real system does not give ideal potential well with infinite barrier height.

In bulk matter means in 3D crystals, for particle's momentum with a quasi-continuous distribution of states, exciton has the wave vector k. Electron-hole pair's propagation is restricted in the plane, in case of quantum wells and could not proceed in the growth direction (here z-axis). Therefore, the quantization of energy states in the growth direction occurs due to the lower dimensionality of the system. In case of 3D crystals, the momentum representation is where i, j, and k are unit vectors along x, y and z direction, respectively. In case of the propagation of 2D particles, momentum expression is $k = k_1 + k_2$ where k_1 is the in-plane and k_2 is the out of plane component of the wave vector. In case of free in-plane propagation, for quasi continuous wave numbers, the transverse momentum is represented by $k_1 = k_x \hat{i} + k_y \hat{j}$. In the longitudinal direction (z direction), the quantized momentum is represented by $k_2 = k_z \hat{z} e_z$, where z is the unit vector along the z direction and e_z is the electronic quantum number.

To design quantum wells, the thickness d of the quantum well is one important parameter. Other parameters in quantum-well systems strongly depend on the value of the thickness. Another important point is the wave-function of charge carriers and the Eigen-energies of the confined particles are modified as a function of d. This can be understood in the following way: When the well is thinner, the wave function will be more squeezed. Therefore the energy levels of the potential well and the amplitude of the wave function will increase. According to the rule of finite potential wells, the standing wave's ends enter into the surrounding material. If the energy difference of the confined charge carriers increases, then the exciton's energy will increase. The effective trapping will be decreased and the thermal outflow of electrons and holes out of the film will be eased when the thickness of the well will be too small. Therefore, to get sufficiently low ground-state energy according to the energy levels, the thickness should be carefully chosen. The oscillator strength of excitons in quantum wells also depends on the thickness and the oscillator strength. Therefore the thickness is an important parameter regarding light–matter interaction. Due to the confinement effect, it comes to significantly higher values than in bulk material [22]. With decreasing the thickness of the well means increasing the confinement of charge carriers, the oscillator strength of the excitonic system increases [23]. Therefore, the collision between the light field and the dipole-like exciton is preferred. It has been seen that this increasing amount is proportional to 1/d according to the numerical calculations [24]. From this discussion, it is possible to understand that the thickness of the quantum well is a very important parameter.

In an ideal case, the exciton Bohr radius of 2D confined excitons is halved compared to the 3D confined excitons, if the potential well is infinitely high. Due to this reason, the binding energy of 2D confined excitons increases by a factor of four compared to the 3D confined excitons [25]. It is never possible to reach this value in case of real quantum well structures. For example, in GaAs, a reduced enhancement factor of around 2 is present [26]. The total energy of 2D excitons can be calculated by:

$$E_X(k,i) = \Delta E_{e,h}(d) - 4E_{bind,i}^{3D} + \frac{\hbar^2 k_1^2}{2m_{X,eff}}$$

(3)

Where, the energy difference of the concerned 2D electron and 3D hole levels is $\Delta E_{e,h}(d)$, the 3D binding energy is $E_{bind,i}^{3D}$ and the effective mass of excitons is $m_{X,eff}$. The values of the 2D Bohr radius and the binding energy can be determined more precisely by taking into account the finite quantum-well thickness [26-28].

Excitation Way

In solids, excitons can be produced by electrical and optical excitation. In the simplest way, the semiconductor crystal absorbs a photon and stimulates an electron then the electron goes from the valence band to the unoccupied state in the conduction band and makes a hole in the valence band (the empty place of an an electron). If the photon energy is equal to the band gap energy, then it is called resonant optical excitation. Whereas, when the photon energy is more than the band gap energy, then it is called non-resonant optical excitation. In this case, the excited electron and hole emit excess energy by phonon emission (see Fig. **3.2 A**) and relax to become hot excitons (k>>0). With increasing time, hot excitons relax (see Fig. 3.2 B) to a Boltzmann distribution of occupied space.

In case of electrical pumping, to get an exciton, a forward bias diode has to be taken in order to flow the electrons from the n-doped side and holes from the p-doped side. If electrical injection is not provided, then a non-resonant pumping scheme can be used to inject a large number of hot charge carriers into the system. In case of reverse bias, a depletion region is generated surrounding the junction and therefore, no current flows through the junction. In this case, the bands can be tilted that allows the quantum-confined Stark effect (QCSE) for research and applications.

The excited excitons can decay either non-radiatively or by radiative recombination by emitting a photon spontaneously within the time period of their lifespan. In quantum-well, the lifetime of the exciton *i.e.* the mean spontaneous recombination time strongly depends on the structural quality and material system [29, 30].

When excitation density is high above the so-called Mott density, then the binding energy decreases due to the increased electronic screening until a Mott transition occurs. One notable point is that in the low-density region, excitons can be treated as good bosons if the fermionic nature of their components is minimal. In the presence of extra free carriers, positively or negatively charged excitons can be formed. Due to the three-particle nature, these are called trions. In this case, one hole and two electrons are bound together or *vice versa*.

In the intermediate density region and at low temperatures, biexcitons are created. Four-particle system's energy is decreased by the corresponding biexciton binding energy due to molecular formation. After its radiative lifespan, a photon is released from the breakdown biexciton state and a single exciton is left behind. Due to the weak binding energy at higher temperatures, biexcitons get typically ionized or detached *via* thermal excitation. Auger effect and exciton–exciton annihilation are also important features. The exciton lifetime is reduced at higher densities due to these effects.

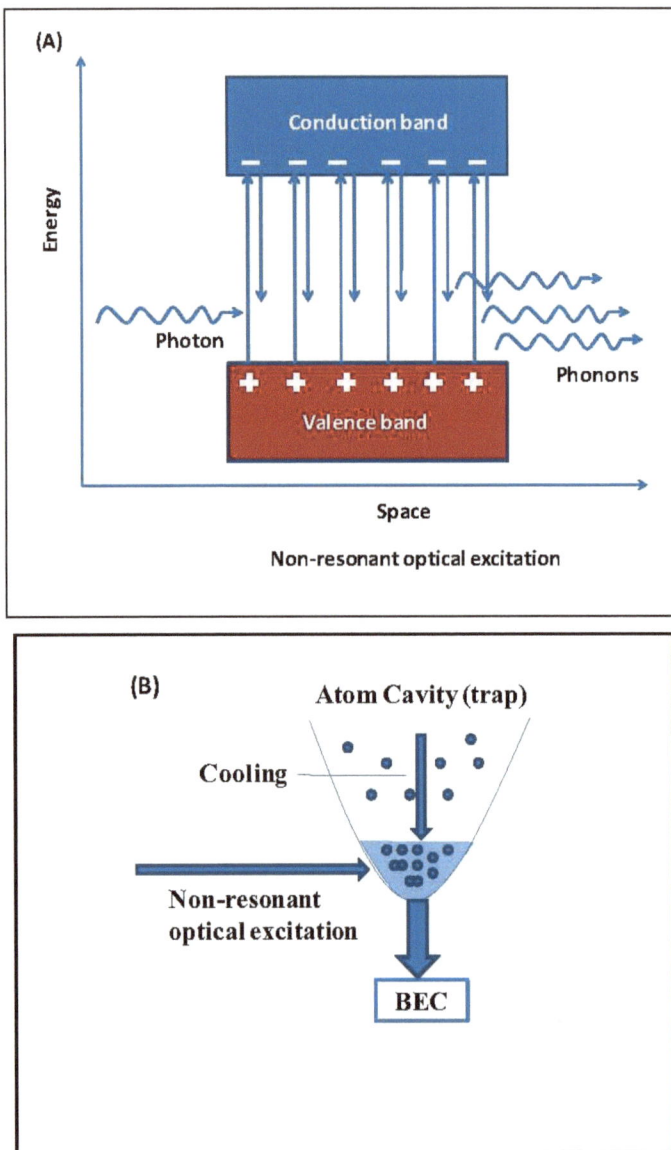

Fig. (3.2). (A) Non resonant optical excitation. Here semiconductor crystal absorbs photon and stimulates electrons in the valence band then the electrons (-) go to the unoccupied state in the conduction band and makes holes (+) in the valence band. These free electrons and holes bound together and form excitons. In case of non-resonant optical excitation, the photon energy is more than the band gap energy. Excited electron and hole pairs relax down by emitting excess energy in the form of phonon emission. **(B)** Cooling process to form Bose Einstein Condensation (BEC). Here by non-resonant excitation, excitons are produced within a potential trap in a semiconductor crystal. After this, excitons relax down to the bottom of the trap and form BEC.

Photons Trapped in Microcavities

An optical resonator is known as a micro cavity when its size is close to the wavelength of the injected laser. On a small scale, a distributed Bragg reflector (DBR) can be used for the trapping of light effectively. The DBR is made by layers of alternating high and low refractive index materials. The optical thickness of each layer is $\lambda/4$. Therefore, the lights reflected from each interface combine in a destructive manner and make it a high-reflectance mirror. When two DBRs are set facing each other with a distance of several micrometers then a microcavity forms.

The trapping ability of a microcavity is defined by the quality factor (Q factor). The Q-factor is represented as:

$$Q = \frac{\omega_c}{\delta\omega_c}$$

(4)

Here ω_c represents the resonant cavity frequency and $\delta\omega_c$ represents the line width of the cavity mode. The Q-factor is the rate of the decay of optical energy within the cavity (from absorption, scattering or leakage through the imperfect mirrors). The fraction of energy lost in a single round-trip around the cavity is represented by Q^{-1}. The Q-factor can exceed 10^5 [31]. The exponentially decaying photon number has a lifetime given by $\tau = \frac{Q}{\omega_c}$. The picture of a cavity layer inserted between two distributed DBRs is given in Fig. **(3.1 B)**. A planar cavity is invariant under in-plane (x-y plane) translations. The in-plane wave vector k has a good quantum number for the free photon dynamics that can be represented by a Hamiltonian of the form [32]

$$H_p = \sum_k E_p(k)a_k^\dagger a_k$$

(5)

Where, k = (kx; ky) is the in-plane momentum. The creation and annihilation operator of photons are represented by $a_k^\dagger$ and a_k respectively. The excitation energy is Ep(k). Here the polarization degrees of freedom of photons are neglected. a_k and $a_k^\dagger$ satisfy the standard Bose commutation rules:

$$[a_k, a_{k'}] = 0$$

$$[a_k, a_{k'}^\dagger] = \delta_{k,k'}$$

(6)

Taking the Fourier transformations of a_k and $a_k^\dagger$, the field operators of photons is obtained. Some photons are uncoupled to excitons in microcavities. Interactions between those photons are weak and usually neglected.

Light-Matter Coupling

When photon goes through the vacuum, it behaves as a particle of the electromagnetic field which oscillates with angular frequency $\omega = 2\pi\nu$, where ν is the frequency. The energy of the photon is $E = \hbar\omega = h\nu = hc/\lambda$ and the momentum is $k = (2\pi/\lambda)\hat{k}$. The relation between angular frequency and momentum is known as dispersion relation, which is represented as $\omega(k) = c|k|$ then $E = \hbar\omega(k) = \hbar c|k|$, where c is the speed of light in vacuum and is a constant. At k=0, $\omega(0) = 0$, therefore the dispersion is minimum. When photons or light goes through a

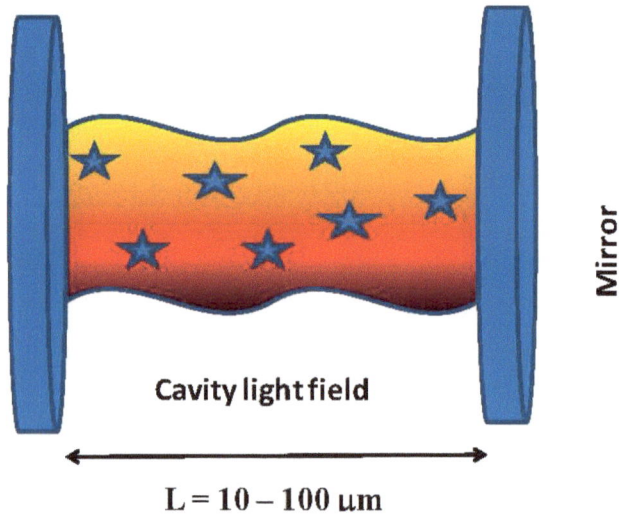

Cavity light field

$\text{L} = 10 - 100 \ \mu\text{m}$

Mirror

(Fig. 3.3) contd.....

(B)

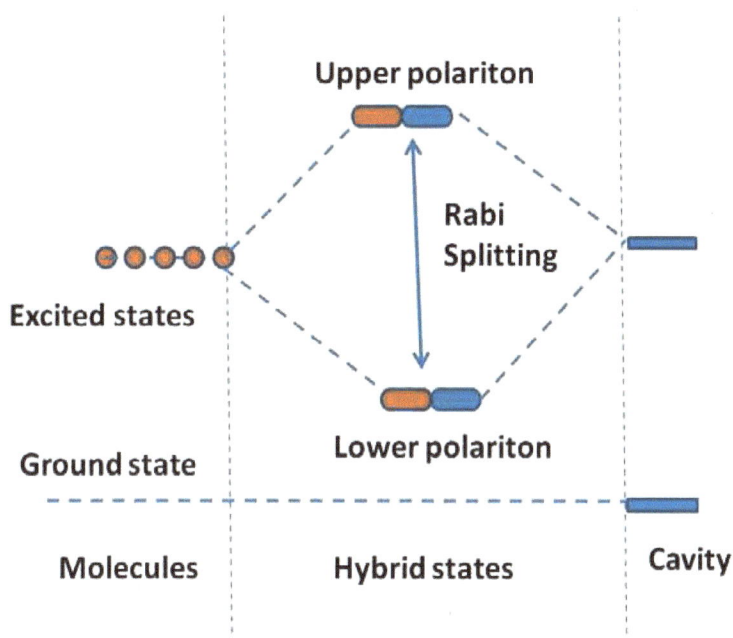

Fig. (3.3). (A) Light matter coupling in the cavity light field. An optical cavity has been created by using two reflective mirrors. A local trapped field of light has been created within this cavity. The wavelength of the light is determined by the distance between the two mirrors. To get polaritons, molecules are placed within this optical cavity. Polaritons are produced by the strong interaction between light and molecules. (B) Strong coupling between the molecules's excited state and cavity's light field and form two new hybrid polariton states (upper polariton state and lower polariton state). When the vibrational frequency of a molecular transition matches the resonant frequency of a cavity mode, the two states hybridize and split into two new states, upper polariton state and lower polariton state by Rabi splitting.

Polarizable medium, then it interacts with matter (Fig. **3.3**). Two things can happen in this situation: first weak coupling, where the excitation of matter and the electromagnetic field maintain their initial properties in the presence of each other and can be considered independent of each other. If the photon resonates with an electronic oscillator in the medium, then the photon can cause an excitation of the matter. Second, light transmits with a modified velocity as in vacuum when the photon is far away from any resonance of a medium.

Polaritons have the wavelike nature of light while maintaining local molecular interactions and structure. Polaritons can be produced by placing molecules within an optical cavity constructed from two reflective mirrors. A local trapped field of light has been created within this cavity, whose wavelength is determined by the distance between the mirrors. In a specific molecular excitation, when the cavity mode is tuned to coincide with its energy, the molecule's excited state and the cavity's light field are coupled strongly and form two new hybrid polariton states. Strong light-matter coupling can occur in the dark, as molecules interact with fluctuations of the quantum vacuum field, a common background of light that persists even in seemingly empty space. Polaritons can therefore be created without pumping the cavity with external light.

In case of dipole approximation, light fields produce an oscillating polarization by interacting with optically allowed transitions through the dipole operator of matter resonances. Electromagnetic radiation emitted from the oscillating polarization superposes with the initial incident wave. Then polariton forms in the strong light-matter interaction region. The quantized mixed state of polarization and light wave interchange energy with the environment like a quasi-particle by integer multiples of the corresponding polariton energies $\hbar\omega$. This coupling of the polarization with the photon generates polariton. In bulk systems, polaritons are in stationary states but at the surface, polaritons transform into photons.

For bulk systems, the dispersion relation ω (k) for polaritons is

$$\frac{c^2 k^2}{\omega^2} = \varepsilon(\omega)$$

(7)

Whereas the so-called polariton equation is obtained by the relation $k^2 = n^2 k_{vaccum}^2 = \varepsilon(\omega)(2\pi / \lambda)^2 = \varepsilon(\omega)(\omega / c)^2$. According to the Lorentz model, the usual behaviour of the dielectric function $\varepsilon(\omega)$ in the neighborhood of a single resonance is given by the equation shown below.

$$\frac{c^2 k^2}{\omega^2} = \varepsilon(b) + \frac{f}{\omega_0^2 - \omega^2 + i\omega\gamma}$$

(8)

Whereas the frequency at the singularity of $\varepsilon(\omega)$ is ω_0, the resonance's oscillator strength is f, the dephasing rate is γ and $\varepsilon(b)$ is the background dielectric constant.

In case of zero damping, the dispersion relation gives two branches; one is the lower branch that comes from the lower energy side and behaves like light close to the resonance before being merged into the energy dispersion of the matter state that increases quasi-particle momentum. Upper branch also behaves like light above the resonance at a low momentum for higher energies.

In case of finite damping, the dispersion relation gets modified due to the absorption in the medium and a reverse back of the lower branch to the upper branch takes place, which reduces reflectivity. Though the event of strong light-matter coupling is not limited to the crystals or solids, but behaviors are important to classify the semiconductor materials.

SPECIAL CONDENSATE FEATURES

Polariton Lasing

Polariton laser is a type of laser that is produced from the Bose condensates of exciton polaritons in semiconductors due to the coherent nature of Bose condensates. First in 1996, Imamoglu *et al.* suggested the concept of laser which is closely associated to the Bose Einstein Condensation of atoms. When a huge number of polaritons macroscopically occupy its lowest quantum state *via* stimulated scattering and form a condensate, then this condensate of polaritons release coherent light that is known as polariton laser (see Fig. **3.4**). According to its principle, a polariton-laser gives a more energy-efficient laser operation. In case of polariton laser, the semiconductor structure is made by an optical microcavity within the distributed Bragg reflectors.

In 2003, H. Deng *et al.* at Stanford University gave an early evidence of polaritonic lasing and a comparison to conventional lasing under optical excitation [33]. After that in 2006, Kasprzak *et al.* [34] fully linked the Polaritonic condensation to the dynamical Bose Einstein condensation. In 2013, the investigators from the University of Michigan [34] and from the University of Würzburg [35] jointly work and showed that electrical pumping of a polariton laser is very essential for the practical use of polaritonic light sources. In 2007, the study on an optically pumped polariton laser at room temperature was illustrated [36, 37].

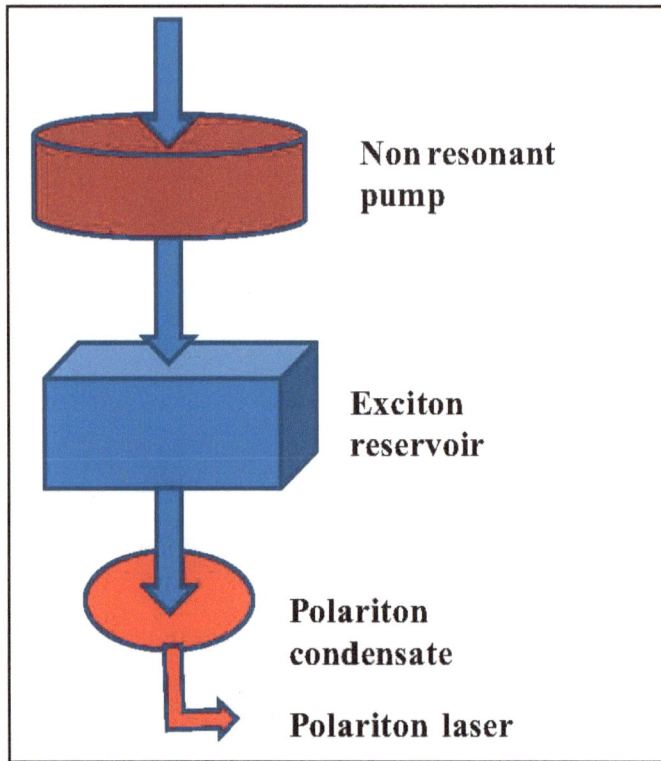

Fig. (3.4). The working process of making polariton laser. Here polaritons are produced within the exciton reservoir by non-resonant pumping. Then, these polaritons relax down and form polariton condensate. The condensate of polaritons finally emits coherent emission of light that is called polariton laser.

In case of a polariton laser, a stimulated scattering process is used to generate coherent radiation. Stimulated scattering process helps to get effective particle loss towards a phase-coherent polariton gas. Due to the apparent number of polaritons in the ground-state, coherent radiation is produced from the bosonic quantum degeneracy of polaritons. Now, due to the polaritonic decay, a beam of coherent photons is released. To make up the optical losses, the dynamical condensate state is constantly repopulated by optical pumping or electrical current injection. Electrically-driven polariton lasers are more acceptable than optically-pumped lasers. Because of the practical uses, electrical pumping has the ability to provide condensates on demand on a device level and can be used to incorporate it into on-chip architecture and implement it in the polaritonic devices within electro-optical systems. One important point is that a polariton laser works on one device with three operative techniques. These are incoherent polariton light emission, photonic lasing, and polariton lasing. In case of population inversion, photonic lasing occurs above the particle densities.

Polariton lasing works below the break-up density and above a critical particle density of strong coupling called Mott transition.

In the active medium semiconductor, lasers contain a non-equilibrium state of the free electron and hole. From the relation between gain and losses in this system, the actual laser threshold is calculated. The actual laser threshold coincides with the Mott density in a high-quality resonator. Using quantum emitters like quantum dots and high-quality microcavities with low mode-volumes, it is possible to get lasers at ultra-low pump-threshold values. Generally, lasing in microcavities is known as lasing in the weak coupling regime. The charge-carrier densities needed for the weak-coupling lasers are much higher than that of the natural polariton laser.

In the strong-coupling region, the threshold less lasing can be obtained. In case of polariton lasers, the particle density required to overcome the critical density for stimulated scattering is much smaller than the general microcavity laser threshold. For this reason, polariton lasers are an attractive source of coherent light with ultra-low operation threshold.

It is very difficult to differentiate polaritonic lasing from conventional (photonic) lasing because both have similar emission properties. A research group from Michigan led by Pallab Bhattacharya did an experimental work by using an external magnetic field to boost the polariton-phonon scattering and the exciton-polariton saturation density. In the active region, they used modulation doping of the quantum wells to increase the polariton-electron scattering. From this investigation, the researchers are able to produce low polariton lasing threshold of 12 A/cm^2 [35]. Another research team from Würzburg, Germany with international collaboration did some crucial experiments in a magnetic field [36]. First-time experimental illustration of an electrically pumped polariton laser has been given by the research team of S Hofling [36]. On June 5, 2014, researchers of Scientist Bhattacharya's team were able to produce the first polariton laser that works at room temperature and is fueled by an electric current [39].

Super Fluidity

Superfluidity can be explained as fluids flowing with infinite thermal conductivity, infinite superconductivity, and zero viscosity. It is a collective and usual nature seen in different systems like liquid Helium and ultracold atomic gasses [40].

The most common principle for superfluidity is the Landau principle. But now it is understood that obeying the Landau principle is not enough for superfluidity. In a superfluid motion, all particles go coherently. Therefore, phase coherence is required to settle over the whole volume. The basis of the modern understanding of superfluidity is the concept of phase coherence and long-range order [41].

It is difficult to analyze the experimental results to test the Landau criterion. When a superfluid gets into in a small tube or a large object, then a superfluid unsteady flow arises its inside and vortices can be excited. It is not possible to measure the critical velocity from this type of experiment because creation and movement of vortices cause the loss of energy. We always get a smaller value of critical velocity than the exact value.

When a superfluid is put into a rotation, then simpler manifestations of superfluidity occur [41]. We first assume that the superfluid parts can be explained by an order parameter. Obtaining the theoretical basis of this hypothesis is not so difficult. It has been proved that at T= 0, He II has a condensate fraction ~10%. But it has also been seen that superfluidity exists without BEC. Therefore, based on the current experimental results, one can only argue that the phase coherence established across 3D superfluids allows the description of the superfluid component by a complex order parameter $|\psi(r)|$. Here $|\psi(r)|$ represents the superfluid density but not condensate density.

$$\psi(r) = |\psi(r)| \exp|i\phi(r)| \tag{9}$$

Here $|\psi(r)|$ is the amplitude and the phase is $|\phi(r)|$. Now, we can describe the superfluid density $\rho_s(r)$ and current density $j_s(r)$ in terms of probability density and current density like quantum mechanical definition by $\rho_s(r) = |\psi(r)|^2$

$$j_s(r) = -\frac{i\hbar}{2m^*} \psi^*(r)\nabla\psi(r) + C = |\psi(r)|^2 \frac{\hbar}{m^*}\nabla\phi(r) \tag{10}$$

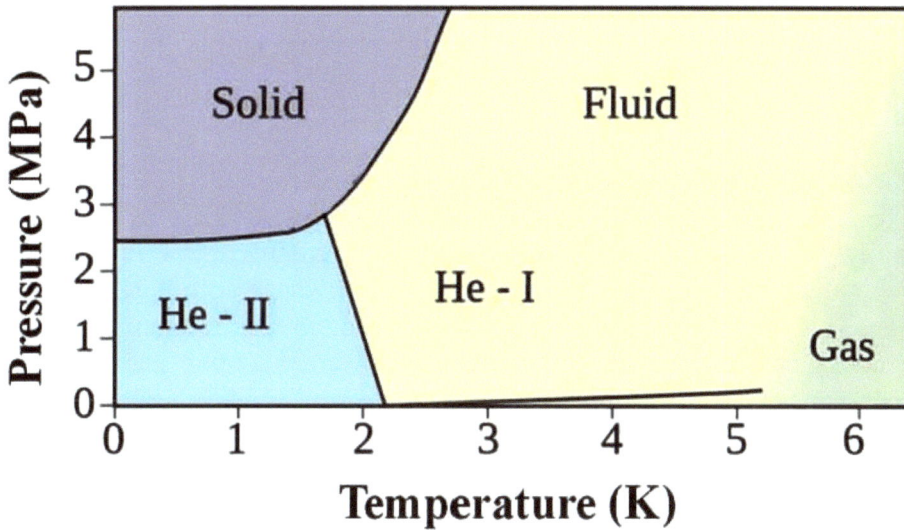

Fig. (3.5). Phase diagram of Helium-4 (⁴He) at low temperatures. When liquid helium-4 is cooled to below 2.17 K (270.98 °C), it becomes a superfluid, that are very unlike in the case of an ordinary liquid. ⁴He remains liquid (sky blue portion) at zero temperature if the pressure is lower than 2.5 MPa (~25 atm).

$j_s(r)/\rho_s(r)$ is the superfluid velocity and it depends on the superfluid phase $v_s = \dfrac{\hbar}{m^*}\nabla\phi(r)$. Wave vector $\psi(r)$ is non zero, so that $\phi(r)$ is well-defined and we get $\nabla\times v_s(r) = 0$. This means the integral of $v_s(r)$ over a close shape of the outer surface will be zero unless this close shape includes one more singularities, which means the points where $\psi(r)$ is disappeared. Since $\phi(r)$ is described as the modulo of 2π, the change of phase $\Delta\phi$ around a closed surface can only be an integer multiple of 2π.

$$\Delta\phi = \oint \nabla\phi.dl = 2\pi l, \qquad l = 0, \pm1, \pm2, \dots\dots$$

$$(10)$$

So, the circulation of Γ around a closed surface is quantized.

$$\Gamma = \oint v_s.dl = l\frac{\hbar}{m^*}$$

$$(11)$$

This equation is known as Onsager–Feynman quantization condition. (Fig. **3.5**) shows a phase diagram of Helium-4 (^{4}He) at low temperatures.

Quantized vortices

Vortices in Polariton Condensates

Generally, any stream of fluid with closed smooth running lines is known as vortex flow. A rotational motion of the fluid around the vortex core is characterized as classical vortices which are easily trackable through the minimum density. The fluid motion can be described by a quantity called circulation that can takes any value. In the case of quantum fluids, vortices are similar like classical vortices but in this case, the circulation is quantized. The quantization of the circulation is directly measured by the phase of the quantum fluid. When going around the vortex core, it linearly varies from 0 to multiple integers of 2π. The vortex charge is calculated by the integer number of 2π of the phase jumps.

The quantized vortices must be present in superfluids such as Bose Einstein condensates. In case of atom condensates and liquid 4He, one experiment has been performed by exciting the fluid with lasers or by rotation of the magnetic trap [42]. In this case, vortices have been detected in several forms like single vortex or vortex lattices. The non-equilibrium nature of the quantum fluid and the disorder in the sample produce vortices. The properties of the polariton quantum fluid depend on the emitted photons. Therefore, it is possible to see the phase winding and the minimum density of available vortices by studying the luminescence image. It is difficult to study the phase winding but the study of minimum density is easy. It is necessary to carry out an interferometric measurement to study the phase winding property.

From the previous studies it is already understood that the interference of two slanted constant phase wave fronts will produce a series of straight interference fringes. Whereas, when constant phase wave fronts interact with a field carrying a vortex *i.e.* a 2π singularity, the outcome will be similar to the constant phase interference. The dissimilarity between the two is the singularity at the center. The interference fringes will show a forklike displacement. The number of fringes changes by one from above below the singularity. A vortex in a polariton condensate is present like a fork-like displacement shape in the interference pattern.

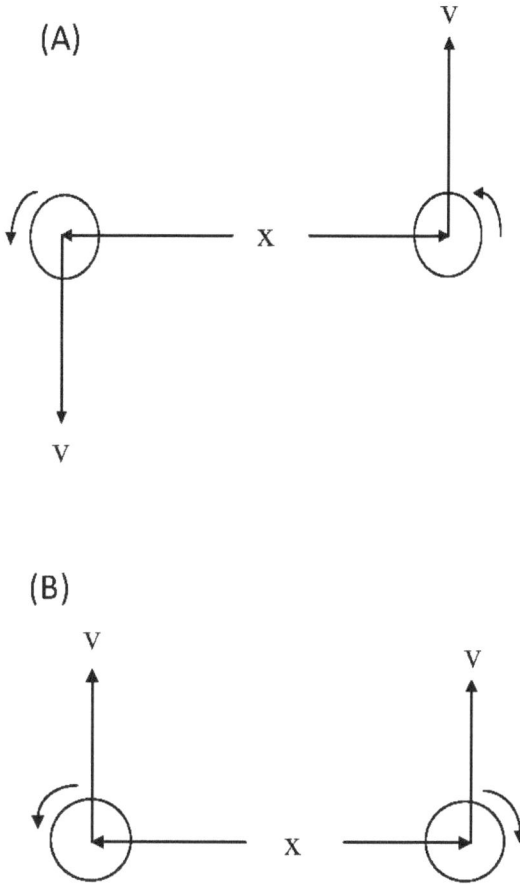

Fig. (3.6). Fig. of the motion of vortex pairs: **(A)** a pair of two vortices rotates. Vortex is a region within a fluid where the flow revolves around an axis line. **(B)** Figure of the motion of vortex anti vortex pair. Anti-vortex is a vortex that rotates in an opposite direction to another that it is paired with. In the vortex-antivortex pair, the vortex is dragged by the velocity field around the antivortex and *vice versa*. As a result, the pair undergoes linear motion.

Vortices may be created in different ways. They can be produced during the very beginning of the condensate creation when the particles with different phases combine to make a large condensate with a common phase. The formation of vortices from gain-loss mechanisms is another possibility. It is possible when the polariton condensates are far out of equilibrium. From the work of reference [42], it has been seen that the nucleation and trapping of vortices may occur. Hydrodynamic nucleation of vortices is another possibility. It forms when a polariton condensate flows with high velocity in opposition to a barrier [43] *i.e.* the motion of particles below a steep potential wall in opposition to a small-size local maximum potential. Pinning of the produced vortices can take place at a place where the motion of vortex is energetically favored. This can also describe the required pinning of the sign of the vortices.

Dynamics of Vortices in Polariton Condensates

Generally in most of the experiments, the condensate is produced non-resonantly and a quasi-continuous wave (CW) laser is used for stimulation. The energy of the quasi-CW laser is much higher than the polariton at resonance. At a high energy, the laser photons make free electrons and holes. After creation, the electron hole pairs move down to the bottom of their respective band. Then in the non-radiative part of the Brillouin zone, free electron hole pairs bound together and form excitons. This non-radiative part of the Brillouin zone is known as reservoir. When exciton density is large enough, they accumulate in the reservoir, then the exciton–exciton collision permits them to scatter to the bottom of the lower polariton branch. The number of polaritons in the lower branch may increase at sufficient excitation density. This lower branch is populated very efficiently by the stimulated scattering.

The authors of reference [44] did an experiment by using a non-resonant femtosecond pulsed excitation. From this work, it has been seen that the shorter is the time to reach this threshold, the larger is the excitation density.

Now to understand the vortex formation in the disorder landscape, one can go through the reference [44]. The researcher of this work observed a forklike displacement in the time-integrated interferogram that is not pinned in the beginning. They show a smooth motion along a well-defined path towards the center of the condensate. Within the first 35 ps of the condensate formation, the most interesting relocation process starts.

Quantized Vortices

One of the most uncommon properties of superfluids is the formation of quantized vortices [45]. According to the discussion before, the superfluid motion will be nonzero if a singularity (a point in which the order parameter $\psi(r)$ vanishes) remains in the fluid. In a three dimensional case, singularities form lines. Some lines discontinued at the boundaries of the fluid. Those are called vortex lines. Some lines stay around themselves. These are called vortex rings. The time independent GPE is:

$$\mu\psi(r) = \left[-\frac{\hbar^2}{2m^*}\nabla^2 + V_{ext}(r) + g|\psi(r)|^2 \right]\psi(r)$$

(12)

Using the time independent GPE (see equation 12), the order parameter around a vortex can be estimated [46]. In the GPE, the single-particle mass is m^*, externally applied potential is $V_{ext}(r,t)$, and the interaction parameter is g. Let us consider an infinite superfluid that contains vortex line along the z-axis. In a loop surrounding the z-axis, the phase changes is an integer multiple of 2π. Therefore in cylindrical coordinates, the order parameter is given by:

$$\psi(r) = f(r)e^{il\phi}$$

(13)

Here, $\psi(r)$ does not depend on z as we assume that the translational symmetry is along the z-axis. By setting the external potential to zero and inserting the value of $\psi(r)$ into equation (12), we get,

$$-\frac{\hbar^2}{2m^*}\frac{1}{r}\frac{\partial}{\partial r}\left(r\frac{\partial f}{\partial r}\right) + \frac{\hbar^2}{2m^*r^2}l^2f + gf^3 = \mu f$$

(14)

If the value of r is very large, it means very far from the latex, we can neglect the term $\frac{\partial f}{\partial r}$ and $\frac{1}{r}$ then, f will be

$$f(r)_{r\to\infty} = \sqrt{\mu/g} \simeq f_0$$

(15)

This equation can be changed to a dimensionless equation by dividing r with the characteristic length or healing length $\zeta = \dfrac{\hbar}{\sqrt{2m^*n_0g}}$ and f with f_0. Where, $n_0(r) = |\psi(r)|^2$ then, it can be solved numerically. A good approximation to the solution of singly quantized vortex (l=1) is

$$f(r) = f_0\frac{r}{\sqrt{2\zeta^2 + r^2}} \Rightarrow \psi(r,\phi) = f_0\frac{r}{\sqrt{2\zeta^2 + r^2}}e^{i\phi}$$

(16)

Now at r=0, the order parameter drops to zero on the z-axis means along the vortex line. The line of singularities is along the z-axis because the phase is unspecified along the z-axis. The magnitude of the order parameter ψ is perturbed in a cylindrical region or radius $\sim \zeta$ surrounding the vortex line. This zone is known as the vortex core. The phase of ψ is perturbed all over the space. Within the closed surface surrounding the vortex line, the change of phase will always be 2π, even if the closed surface is infinitely far from the vortex core. These qualitative features are always true even for the system where GPE is not a good approximation, like in a strongly interacting superfluid, He II.

A vortex line is a metastable entity. Any vortices do not stay in the ground state of a superfluid. The disappearance of a vortex from the inside of the fluid is highly uncertain since the phase is perturbed all over surrounding a vortex line. A vortex can only disappear by destroying it with an antivortex of opposite circulation or by coming to the boundary. The energy per unit length of a vortex line can be estimated by the integration of the solution (16). Another easy and more straightforward way [45] to do vortex calculation is using Onsager–Feynman quantization condition (equation 11). Around the vortex line the velocity field is:

$$v(r) = \frac{\Gamma}{2\pi r} \tag{17}$$

The kinetic energy per unit length correlated with a single vortex is given by

$$K = \int_\zeta^R \frac{1}{2} \rho_s \left(\frac{\Gamma}{2\pi r} \right)^2 2\pi r dr = \frac{\rho_s \Gamma^2}{4\pi} \ln\left(\frac{R}{\zeta} \right) \tag{18}$$

Where the core of the vortex has been neglected and it has been assumed that the fluid lengthens up to radius R. Γ is the whole circulation lh/m^*, where $l = 0, \pm 1, \pm 2, \ldots\ldots$. In a 3D case, the sum of the energy increases linearly with the system size because the vortex line is in proportion to its length. On the other hand, in a two dimensional case, the vortex energy excites easily because in this case, vortex energy rises with the logarithm of the system size.

Now if we consider a multiple quantized vortex (l > 1), then from equation (18) it can be predicted that the vortex state will finally detach into multiple single quantized vortices. According to equation (18), the kinetic energy of a double quantized vortex (l = 2) is four times the energy of a single quantized vortex.

Therefore it is strongly desirable that double quantized vortex splits into two singly quantized vortices. The vortices can change the flow characteristics of the fluid, and for this reason, they are very important in superfluidity. In case of atomic gasses, the theory of quantized vortices has been discussed in references [47, 48] and all relevant experiments have been discussed in reference [49].

Now if we consider a system of one vortex line and one antivortex line and assume that the lines are straight and parallel to each other at a distance x, then a good approximation for the energy of this system is defined as:

$$K = 2\int_\zeta^x \frac{1}{2}\rho_s \left(\frac{\Gamma}{2\pi r}\right)^2 2\pi r dr = \frac{\rho_s \Gamma^2}{2\pi}\ln\left(\frac{x}{\zeta}\right)$$

(19)

The force by which the vortex and antivortex will attract with each other is shown as:

$$F(x) = -\frac{dK}{dx} = -\frac{\rho_s \Gamma^2}{2\pi x}$$

(20)

Due to Coulomb's law of attraction, the complication of an ensemble of vortices and antivortices in a plane can be decreased in an ensemble of dipole charges. This proposal was given by the scientists Kosterlitz and Thouless. A brief description is given below.

There is a rotating superfluid flow around a vortex. Thus, the flow surrounding the vortex attracts the antivortex and *vice versa*, in a system of a vortex and antivortex. The pair proceeds along the perpendicular direction to the line that joins the vortex and antivortex in the absence of any other forces, as shown in Fig. (**3.6**) (bottom) with velocity $v = \frac{\Gamma}{2\pi x}$. On the other hand, in case of two vortices, the system revolves with angular velocity $\omega = \Gamma / \pi x^2$ (Fig. **3.6** top).

CONCLUSION

Polariton is a very important quasiparticle to get the BEC. It can be created in semiconductors, quantum wells, microcavities, *etc*. Polariton lasing, superfluidity, and quantized vortices is the special condensate feature of polaritons. The BEC of polaritons has potential practical applications. Therefore, the research in this field is very important.

REFERENCES

[1]   Polariton **2022**. Available from: https://en.wikipedia.org/wiki/Polariton [Accessed: Oct. 13, 2022].

[2]   Fox, M. *Optical Properties of Solids,* 2nd ed; Oxford University Press: Oxford, **2010**, p. 107.

[3]   Eradat, N.; Sivachenko, A.Y.; Raikh, M.E.; Vardeny, Z.V.; Zakhidov, A.A.; Baughman, R.H. Evidence for braggoriton excitations in opal photonic crystals infiltrated with highly polarizable dyes. *Appl. Phys. Lett.,* **2002**, *80*(19), 3491-3493.
        http://dx.doi.org/10.1063/1.1479197

[4]   Yuen-Zhou, J.; Saikin, S.K.; Zhu, T.; Onbasli, M.C.; Ross, C.A.; Bulovic, V.; Baldo, M.A. Plexciton dirac points and topological modes. *Nat. Commun.,* **2016**, *7*(1), 11783.
        http://dx.doi.org/10.1038/ncomms11783 PMID: 27278258

[5]   Tonks, L.; Langmuir, I. Oscillations in Ionized Gases. *Phys. Rev.,* **1929**, *33*(2), 195-210.
        http://dx.doi.org/10.1103/PhysRev.33.195

[6]   Tolpygo, K.B. Physical properties of a rock salt lattice made up of deformable ions. *Zh. Eks.Teor. Fiz.,* **1950**, *20*(6), 497-509. [in Russian].

[7]   Tolpygo, K.B. Physical properties of a rock salt lattice made up of deformable ions *Ukr. J. Phys.,* **2008**, *53* English translation

[8]   Huang, K. Lattice vibrations and optical waves in ionic crystals. *Nature,* **1951**, *167*(4254), 779-780.
        http://dx.doi.org/10.1038/167779b0

[9]   Huang, K. On the interaction between the radiation field and ionic crystals. *Proc. R. Soc. Lond. A Math. Phys. Sci.,* **1951**, *208*(1094), 352-365.
        http://dx.doi.org/10.1098/rspa.1951.0166

[10]   Otto, A. Excitation of nonradiative surface plasma waves in silver by the method of frustrated total reflection. *Zeitschrift für Physik A Hadrons and nuclei,* **1968**, *216*(4), 398-410.
        http://dx.doi.org/10.1007/BF01391532

[11]   Lerario, G.; Fieramosca, A.; Barachati, F.; Ballarini, D.; Daskalakis, K.S.; Dominici, L.; De Giorgi, M.; Maier, S.A.; Gigli, G.; Kéna-Cohen, S.; Sanvitto, D.; Sanvitto, D. Room-temperature superfluidity in a polariton condensate. *Nat. Phys.,* **2017**, *13*(9), 837-841.
        http://dx.doi.org/10.1038/nphys4147

[12]   Hignett, K. *Physics creates new form of light that could drive the quantum computing revolution,* **2018**. Available: Newsweek. https://www.newsweek.com/photons-light-physics-808862 [Accessed: Oct. 14, 2022].

[13]   Liang, Q.Y.; Venkatramani, A.V.; Cantu, S.H.; Nicholson, T.L.; Gullans, M.J.; Gorshkov, A.V.; Thompson, J.D.; Chin, C.; Lukin, M.D.; Vuletić, V. Observation of three-photon bound states in a quantum nonlinear medium. *Science,* **2018**, *359*(6377), 783-786.
        http://dx.doi.org/10.1126/science.aao7293 PMID: 29449489

[14]   Klingshirn, C.F. *Semiconductor Optics*; Springer: Berlin, **2012**.
        http://dx.doi.org/10.1007/978-3-642-28362-8

[15]   Bastard, G. *Wave Mechanics Applied to Semiconductor Heterostructures*; Wiley-Interscience: New York, **1991**.

[16]   Ivchenko, E.L. *Optical Spectroscopy of SemiconductorNanostructures*; Springer: Berlin, **2004**.

[17]   Davies, J.H. *The Physics of Low-dimensional Semiconductors: An Introduction*; Cambridge University Press: Cambridge, **1998**.

[18]   Yamamoto, Y.; Tassone, F.; Cao, H. *Semiconductor Cavity Quantum Electrodynamics*; Springer: Berlin, **2000**.

[19]   Yu, P.Y.; Cardona, M. *Fundamentals of Semiconductors: Physics and Materials Properties*; Springer-Verlag: Berlin, Heidelberg, **2010**.
       http://dx.doi.org/10.1007/978-3-642-00710-1

[20]   Harrison, P. *Quantum Wells, Wires and Dots: Theoretical and Computational Physics,* 2nd ed; Wiley-Interscience, **2005**.
       http://dx.doi.org/10.1002/0470010827

[21]   Lagoudakis, K. *The Physics of Exciton-Polariton Condensates*; EPFL Press, **2013**.
       http://dx.doi.org/10.1201/b15531

[22]   Deng, H.; Haug, H.; Yamamoto, Y. Exciton-polariton bose-einstein condensation. *Rev. Mod. Phys.,* **2010**, *82*(2), 1489-1537.
       http://dx.doi.org/10.1103/RevModPhys.82.1489

[23]   Andreani, L.C.; Pasquarello, A. Accurate theory of excitons in GaAs- Ga $1 - x$ Al x As quantum wells. *Phys. Rev. B Condens. Matter,* **1990**, *42*(14), 8928-8938.
       http://dx.doi.org/10.1103/PhysRevB.42.8928 PMID: 9995104

[24]   Masumoto, Y.; Matsuura, M.; Tarucha, S.; Okamoto, H. Direct experimental observation of two-dimensional shrinkage of the exciton wave function in quantum wells. *Phys. Rev. B Condens. Matter,* **1985**, *32*(6), 4275-4278.
       http://dx.doi.org/10.1103/PhysRevB.32.4275 PMID: 9937602

[25]   Kavokin, A.V.; Baumberg, J.J.; Malpuech, G.; Laussy, F.P. *Microcavities*; Oxford University Press: Oxford, **2007**.
       http://dx.doi.org/10.1093/acprof:oso/9780199228942.001.0001

[26]   Leavitt, R.P.; Little, J.W. Simple method for calculating exciton binding energies in quantum-confined semiconductor structures. *Phys. Rev. B Condens. Matter,* **1990**, *42*(18), 11774-11783.
       http://dx.doi.org/10.1103/PhysRevB.42.11774 PMID: 9995485

[27]   Greene, R.L.; Bajaj, K.K.; Phelps, D.E. Energy levels of Wannier excitons in G a A s $-$ Ga $1 - x$ Al x As quantum-well structures. *Phys. Rev. B Condens. Matter,* **1984**, *29*(4), 1807-1812.
       http://dx.doi.org/10.1103/PhysRevB.29.1807

[28]   Priester, C.; Allan, G.; Lannoo, M. Wannier excitons in GaAs- Ga $_1$ $_x$ Al $_x$ As quantum-well structures: Influence of the effective-mass mismatch. *Phys. Rev. B Condens. Matter,* **1984**, *30*(12), 7302-7304.
       http://dx.doi.org/10.1103/PhysRevB.30.7302

[29]   Dreybrodt, J.; Forchel, A.; Reithmaier, J.P. Excitonic transitions in GaInAs/GaAs surface quantum wells *J. Phys.*, **1993**, *IV 3(C5)*, , 265-268.
       http://dx.doi.org/10.1051/jp4:1993552

[30]   Dang, L.S.; Heger, D.; André, R.; Bœuf, F.; Romestain, R. Stimulation of polariton photoluminescence in semiconductor microcavity. *Phys. Rev. Lett.*, **1998**, *81*(18), 3920-3923.
       http://dx.doi.org/10.1103/PhysRevLett.81.3920

[31]   Reitzenstein, S.; Hofmann, C.; Gorbunov, A.; Strauß, M.; Kwon, S.H.; Schneider, C.; Löffler, A.; Höfling, S.; Kamp, M.; Forchel, A. Al As/Ga As micropillar cavities with quality factors exceeding 150.000. *Appl. Phys. Lett.*, **2007**, *90*(25), 251109.
       http://dx.doi.org/10.1063/1.2749862

[32]   Carusotto, I.; Ciuti, C. Quantum fluids of light. *Rev. Mod. Phys.*, **2013**, *85*(1), 299-366.
       http://dx.doi.org/10.1103/RevModPhys.85.299

[33]   Deng, H.; Weihs, G.; Snoke, D.; Bloch, J.; Yamamoto, Y. Polariton lasing vs. Photon lasing in a semiconductor microcavity. *Proc. Natl. Acad. Sci. USA,* **2003**, *100*(26), 15318-15323.
       http://dx.doi.org/10.1073/pnas.2634328100 PMID: 14673089

[34]   Kasprzak, J.; Richard, M.; Kundermann, S.; Baas, A.; Jeambrun, P.; Keeling, J.M.J.; Marchetti, F.M.; Szymańska, M.H.; André, R.; Staehli, J.L.; Savona, V.; Littlewood, P.B.; Deveaud, B.; Dang, L.S. Bose–Einstein condensation of exciton polaritons. *Nature,* **2006**, *443*(7110), 409-414.
       http://dx.doi.org/10.1038/nature05131 PMID: 17006506

[35]   Bhattacharya, P.; Xiao, B.; Das, A.; Bhowmick, S.; Heo, J. Solid state electrically injected exciton-polariton laser. *Phys. Rev. Lett.*, **2013**, *110*(20), 206403.
       http://dx.doi.org/10.1103/PhysRevLett.110.206403 PMID: 25167434

[36]   Schneider, C.; Rahimi-Iman, A.; Kim, N.Y.; Fischer, J.; Savenko, I.G.; Amthor, M.; Lermer, M.; Wolf, A.; Worschech, L.; Kulakovskii, V.D.; Shelykh, I.A.; Kamp, M.; Reitzenstein, S.; Forchel, A.; Yamamoto, Y.; Höfling, S. An electrically pumped polariton laser. *Nature,* **2013**, *497*(7449), 348-352.
       http://dx.doi.org/10.1038/nature12036 PMID: 23676752

[37]   Christopoulos, S.; von Högersthal, G.B.H.; Grundy, A.J.D.; Lagoudakis, P.G.; Kavokin, A.V.; Baumberg, J.J.; Christmann, G.; Butté, R.; Feltin, E.; Carlin, J.F.; Grandjean, N. Room-temperature polariton lasing in semiconductor microcavities. *Phys. Rev. Lett.*, **2007**, *98*(12), 126405.
       http://dx.doi.org/10.1103/PhysRevLett.98.126405 PMID: 17501142

[38]   Johnston, H. Polariton laser reaches room temperature *Physics World,* **2007**. Available from: https://physicsworld.com/a/polariton-laser-reaches-room-temperature/ [Accessed: 23rd December, 2022].

[39]   Bhattacharya, P.; Frost, T.; Deshpande, S.; Baten, M.Z.; Hazari, A.; Das, A. Room temperature electrically injected polariton laser. *Phys. Rev. Lett.*, **2014**, *112*(23), 236802.
       http://dx.doi.org/10.1103/PhysRevLett.112.236802 PMID: 24972222

[40]   Annett, J.F. Superconductivity, superfluids, and condensates.*Oxford master series in condensed matter physics*; Oxford University Press: Oxford, **2004**.

http://dx.doi.org/10.1093/oso/9780198507550.001.0001

[41]   Leggett, A.J. Superfluidity. *Rev. Mod. Phys.,* **1999**, *71*(2), S318-S323.
       http://dx.doi.org/10.1103/RevModPhys.71.S318

[42]   Posazhennikova, A. ***Colloquium*** : Weakly interacting, dilute Bose gases in 2D. *Rev. Mod. Phys.,* **2006**, *78*(4), 1111-1134.
       http://dx.doi.org/10.1103/RevModPhys.78.1111

[43]   Kosterlitz, J.M.; Thouless, D.J. Long range order and metastability in two dimensional solids and superfluids. (Application of dislocation theory). *J. Phys. C Solid State Phys.,* **1972**, *5*(11), L124-L126.
       http://dx.doi.org/10.1088/0022-3719/5/11/002

[44]   Nardin, G.; Lagoudakis, K.G.; Wouters, M.; Richard, M.; Baas, A.; André, R.; Dang, L.S.; Pietka, B.; Deveaud-Plédran, B. Dynamics of long-range ordering in an exciton-polariton condensate. *Phys. Rev. Lett.,* **2009**, *103*(25), 256402.
       http://dx.doi.org/10.1103/PhysRevLett.103.256402 PMID: 20366268

[45]   Donnelly, R.J. *Quantized Vortices in Helium II*; Cambridge University press: Cambridge, **1991**.

[46]   Pethick, C.J.; Smith, H. *Bose-Einstein Condensation in Dilute Gases,* 2nd ed; Cambridge University press: Cambridge, **2008**.
       http://dx.doi.org/10.1017/CBO9780511802850

[47]   Fetter, A.L.; Svidzinsky, A.A. Vortices in a trapped dilute Bose-Einstein condensate. *J. Phys. Condens. Matter,* **2001**, *13*(12), R135-R194.
       http://dx.doi.org/10.1088/0953-8984/13/12/201

[48]   Fetter, A.L. Vortices and dynamics in trapped Bose-Einstein condensates. *J. Low Temp. Phys.,* **2010**, *161*(5-6), 445-459.
       http://dx.doi.org/10.1007/s10909-010-0202-7

[49]   Anderson, B.P. Resource article: Experiments with vortices in superfluid atomic gasses. *J. Low Temp. Phys.,* **2010**, *161*(5-6), 574-602.
       http://dx.doi.org/10.1007/s10909-010-0224-1

CHAPTER 4

Bose Einstein Condensation of Excitons

Abstract: The research on the Bose Einstein Condensation (BEC) of polaritons is of increasing importance due to its many technological applications. Several experimental and theoretical methods have been applied for the investigation on the BEC of excitons. Different experimental procedures and theoretical methods have been discussed in this chapter. Current research progress in this field and the applications of the BEC are also explained.

Keywords: Boltzmann transport equation, Developments of the Bose Einstein Condensation of excitons, Experimental procedures to get Bose Einstein Condensation of excitons, Technological applications of the Bose Einstein Condensation of excitons.

EXPERIMENTAL TECHNIQUES TO GET BEC

Early Experiments

Spectral Analysis Experiment

Till now, many experimental methods have been applied by different researchers to get the BEC of excitons. One of them is the spectral analysis experiment. This is the first experiment which shows the ability to get exciton BEC in Cu_2O [1, 2]. In that experiment, after excitation, a thin shape of excitons like a pancake is created that were not trapped and able to outflow from the surface. For excitation, a green laser is used that was absorbed within 5 mm of the surface. But at high density, exciton stays at the point of formation for a few nanoseconds. The orthoexciton phonon assisted luminescence spectrum could be well fitted by the Bose Einstein distribution:

$$I(E) \propto D(E) f(E) \propto E^{1/2} \frac{1}{e^{(E-\mu)/k_B T} - 1} \tag{1}$$

Where, E is the exciton kinetic energy, $I(E)$ is the intensity of the exciton distribution, $D(E)$ is the density of states of the excitons that is proportional to $E^{1/2}$ in three dimensions and $f(E)$ is the occupation number of the excitons that is proportional to $e^{-E/k_B T}$ at a low density.

With $|\mu / k_B T| \sim 0.1$ stays over a wide range of densities. Therefore, the BEC gas did not cross the phase boundary but stayed near it. This is termed as Bose saturation. The absolute density of the particles is:

$$N = \int_0^\infty \frac{1}{e^{(E-\mu)/k_B T} - 1} D(E) dE$$

(2)

According to equation (2) at the absolute density, the spectrum fits the Bose Einstein distribution of the particles.

At each point in time by taking the ratio of the total luminescence intensity and the measured exciton cloud volume, the relative change in the density can be calculated. Then the densities from the fits have been compared to the relative change in the density. The researchers of reference [1] reported that over a wide range of densities, the variation in the density can be calculated from the fits of those agreed with the variation in the density from the intensity data within a factor of two.

Later the researchers of reference [3] investigated on the BEC of paraexcitons. In Cu_2O by applying uniaxial stress, they reduced the multiplicity of the orthoexciton ground state. It has been claimed that according to the ideal gas theory, the paraexciton density is higher than the density to obtain BEC. The emission intensity of the orthoexciton emission line is 500 times stronger than the paraexcitons emission line. Another brighter emission line also stays near it. Therefore, it is very difficult to make line shape analysis of the paraexcitons phonon assisted emission.

In these above experiments, by taking an image of a three-dimensional (3D) crystal in two dimensions (2D), the photons are collected. Then the fluorescence from the exciton gas must be united over one dimension. Therefore the difficulty arises in interpreting the three-dimensional data integrated over one dimension. To overcome this difficulty, the researchers of reference [4, 5] did their experiments with excitons, or exciton-polaritons, in two dimensions.

Transport Experiments

From the transport experiment, it is possible to know about the transport of the excitons rather than spectral signatures [6-9]. In this experiment, to get low-density exciton gas flow within a three-dimensional crystal of Cu_2O, a red laser of long-wavelength is used. Here the exciton energy nearly resonates with the photon energy of the laser. To create a very concentrated exciton cloud at the surface, a very intense pulse of green light is used. That exciton cloud excites one surface of the crystal. From the opposite side of the crystal, excitons are detected. Sometimes, a bimetal detector is used over a millimeter distance that converts the excitons into free electrons and holes. An external current is produced from these holes and free electrons. If excitons relocate through the crystal into the detector, then the signal is produced.

The researchers of reference [6-9] reported that when the green laser pulse intensity exceeded a critical threshold value, the exciton signal increased sharply on the back side of the crystal. The speed of the sound throughout the crystal and the arrival time of the pulse are similar. The author explained the excitons' motion throughout the crystal as superfluid motion when the exciton density exceeded a critical density threshold value.

But both spectral signature experiments and transport experiments have some drawbacks and because of this, any of these experiments did not show BEC. To get very high exciton density, very intense surface excitation has been used in these two experiments. In both of the cases, the exciton gas was produced very far from the equilibrium. Therefore the outflow of the phonon wind and heat flow of the excitation region should be modeled. Because of these reasons, both experiments were not so successful.

Nowadays Experiments

Experimental Procedure 1

Nowadays, experimental procedures are little different than the early experiments. In 2012, the researchers [10] studied the Bose Einstein condensation of excitons in cuprous oxide at ultra-cold temperatures. For their experiment, they used narrow-band tunable dye laser (Coherent CR599, laser dye Rhodamine 6G). The laser was pumped by a 5W green solid state laser (Verdi 5). A closed feedback loop was used to stabilize the laser power nearly 1%. A wave meter of High Finesse WS7 and resolution 60 MHz was used to measure the laser frequency and line width. In the middle of the spectrometer exit slit and the detector, a fourfold magnification optical imaging system was used to improve the spectral and spatial resolutions. A natural cuprous oxide crystal originally found in Namibia has been used for their experiment in the form of millimeter sized cubic specimens with clear-cut surfaces. They reported that the quality of that crystal is very high. It has low defect density and long paraexciton lifetime up to $1\mu s$.

In this case, the exciton gas was confined within a potential trap. This potential trap has been created by Hertzian stress technique [11-14] (see Fig. **4.1**). In 1882, the scientist Heinrich Hertz solved the contact problem of two elastic bodies with curved surfaces. This is helpful to solve the problems in contact mechanics. When two surfaces are in contact and a load is applied, then a small contact area is formed, and the surfaces experience very high stress. This stress is known as Hertz (or Hertizian) contact stress. In that experiment, a glass-made spherical stressor of radius 7.75 mm is kept on the flat surface of the crystal with a force F along the z direction as they denote. Due to this potential trap, an impound potential is made where the energies of orthoexcitons and paraexcitons are low compared to bulk. Impound potential can be measured from the known parameters of the yellow exciton series or states. The indirect absorption process has been used to create excitons that involve an odd parity optical phonon Γ_3^- into the orthoexcitons in the trap. After creation, orthoexcitons converted into paraexcitons due to their short lifetime. The laser beam was fixed away from the stressor lens about 100 μm far from the trap center, in the positive z-direction. The energy of the laser photons was tuned slightly around 0.5 meV, below the phonon sideband in the bulk at 2048.56 meV to confine the created orthoexcitons in the trap and to keep away from any excitation outside of the trap. If the orthoexciton goes outside the trap, then it may create or lose excitons. About half of the incoming photons are converted into excitons through this process.

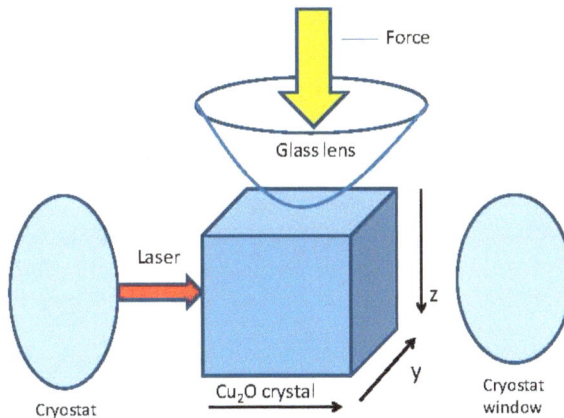

Fig. (4.1). Hertzian stress technique by which the potential trap is created in the experiment. The stress close to the area of contact between two surfaces or two spheres of different radii due to applied load is known as Hertzian contact stress. A glass lens is kept on the Cu_2O crystal and a force is applied on the glass lens then a potential trap is created within the crystal. Now from the left-hand side, a laser beam is inserted into the crystal through the cryostat window and from the right-hand side, the detection is performed by the CCD camera through the cryostat window.

To measure the number of excited orthoexcitons, the **k**-vector of the excitation beam has been taken along the direction of observation. The released light falls onto the entry slit of a high resolution triple spectrograph (T64000, Jobin Yvon) in the experimental setup (Fig. **4.2**) and an image has been taken from that slit. This spectrograph can be used either in the subtractive mode or the additive mode with a diffraction-limited spatial resolution of the order of 10 μm. A cylindrical lens has been placed in front of the entrance slit to correct the *imperfection* of the spectrograph.

The luminescence of width $2\Delta y$ from a small strip centered in the trap to obtain a z resolved spectrum $I(z, \omega)$, along the way of the applied strain. Then it was united along the x-direction that is perpendicular to the z-direction. Detection is carried out either by a nitrogen-cooled charge-coupled device (CCD) camera with a high quantum efficiency, for example Andor Newton or by an intensified CCD camera, for example Andor iStar, that has minimum temporal resolution of 5 ns. Both the spectral and spatial dependencies of the excitonic luminescence can be measured by this process from which the absolute luminescence intensities of the exciton states can be calculated.

The researchers of reference [15] concluded that a Bose Einstein condensate occurs for the paraexcitons in Cu_2O at ultracold temperatures in the span of 100 mK.

Experimental Procedure 2

The researchers [16] performed an experiment to investigate Bose Einstein condensation of excitons at sub kelvin temperatures. For this, a 3He refrigerator is used to cool the crystal as low as 278 mK. They have prepared a three-dimensional harmonic potential trap by using an inhomogeneous strain to gather excitons under quasi-continuous- -wave (CW) feeding and to avoid ballistic diffusion of excitons. To stop the heat on the coldest plate of the refrigerator, the size of the numerical aperture and the optical windows were chosen very carefully. To create the stress along the (100) axis of the crystal, a lens and a built spring were pressed on the crystal. To give a static pressure on the crystal, a sample holder with a built-in spring has been designed to prevent any incoming heat. Weak luminescence light was collected from the (110) surface that was emitted from the trapped paraexciton gas. Compared with the life time of 300 ns, they applied the trap frequencies of 106.76 MHz in the x and y direction and 157 MHz in the z-direction.

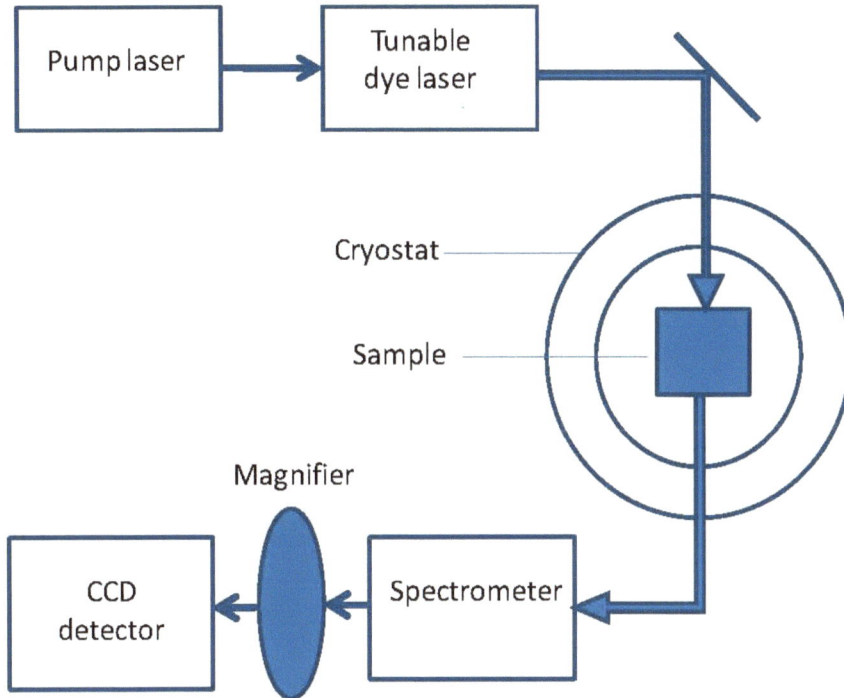

Fig. (4.2). The main items of one experimental setup. A dye laser propagates along x direction and produces excitons. A dye absorbs the pump laser radiation and can direct enough energy of proper wavelength into the dye with a small amount of input energy. First, the dye laser is focused onto a small round area that closes to the trap or directly into the trap. The trap is created by the Hertzian stress technique and the sample is within the trap. A dye laser falls on the sample and produces excitons. Then excitons spread over wide area of the trap and after some time, it relaxes down towards the bottom of the trap. The emission from the trap passes through a spectrometer. The emission out of the trap is observed along the Z direction while integrating along a small strip along the y and the x direction. A lens has been used in front of the spectrometer to correct the *imperfection* of the spectrograph. The detection has been made by a CCD camera.

To generate excitons, they used a CW ring dye excitation laser of 606 nm. By applying a high-isolation switch and an acousto-optic modulator, the excitation beam was chopped for an optical power of more than 250 μW. To reduce the scattered light, this excitation beam was coupled into a single-mode optical fiber. To get the average constant power, the duty cycle (Dcycle) has been set properly. The paraexciton number increases only sub linearly at high excitation intensity. For this reason, in this case high excitation intensity is essential.

The incoherent 1S orthoexcitons are generated near the bottom of the trap *via* a phonon-assisted absorption process in a cigar-shaped volume by the excitation of CW laser. 1s orthoexciton level lies 12 meV above the 1s paraexciton level. Then the orthoexcitons converted into the paraexcitons within the nanosecond range and flowed towards the bottom of the trap. When the paraexcitons went to the bottom of the trap, their kinetic energy emitted in the form of phonon. Cold paraexcitons gather around the minimum potential. Then in the experiment, the spatially resolved and time-integrated luminescence spectra have been detected from the direct radiative recombination of the paraexcitons. This is slightly permitted by the strain field.

The luminescence image of the trapped exciton gas was magnified four times then it was accumulated and spread by a 50 cm spectrometer. To detect the time-integrated signal, an EMCCD camera (Andor, DU970N-BV) was used. The time range of integration was 100 s to 1500 s. Only in the central part, 7.5 μm of the gas is captured by the entrance slit of the spectrometer, parallel to the z-axis which is equivalent to 30 μm in case of a magnified image. In the imaging section of the setup, the vibration of the sample with an amplitude of around 10 μm was canceled using a pair of piezoelectric transducers. The piezoelectric transducers monitor the movement using a laser diode and a position-sensitive photodetector. The spatial resolution of the photodetector was 0.76 μm. Within their allowable integration time, the detection noise level and the intensity of the Γ5 phonon-assisted luminescence of paraexcitons are comparable.

THEORETICAL MODELING TO GET THE RELAXATION KINETICS AND BEC OF EXCITONS

Theoretical Methodology

Condensation kinetics of excitons includes creation, relaxation and decay processes since excitons have lifetime on the order of ns. The semi classical Boltzmann rate equation can be used to explain the kinetics of excitons [17]. The Boltzmann rate equation can be written as:

$$\frac{\partial Q}{\partial t}+\left(\vec{\upsilon}\cdot\overrightarrow{\nabla_r}Q+\vec{F}\cdot\frac{1}{\hbar}\overrightarrow{\nabla_k}Q\right)_{drift\,\&\,force}=\left(\frac{\partial Q}{\partial t}\right)_{collision+interaction} \qquad (3)$$

The Boltzmann equation explains the particle's number $Q(\vec{r},\vec{k},t)$ in terms of radius $\vec{r}$, momentum $\hbar\vec{k}$ and time t. Here the velocity is υ, the force is $\vec{F}$ and the nabla operators in r and k space are ∇_r and ∇_k, respectively. The drift and force terms are the left part of the equation (3) and the collision and interaction terms are the right part of the equation (3). Excitons phonon scattering, excitons-excitons scattering and Auger decay are the collision and interaction terms.

$$\left(\frac{\partial Q}{\partial t}\right)_{coll+int} = \left(\frac{\partial Q}{\partial t}\right)_{exciton\text{-}phonon} + \left(\frac{\partial Q}{\partial t}\right)_{exciton-exciton} + \left(\frac{\partial Q}{\partial t}\right)_{Auger\_decay} \tag{4}$$

Excitons phonon scattering term is given by [17, 9]

$$
\begin{aligned}
\left(\frac{\partial Q_{\vec{k}}}{\partial t}\right)_{polariton\ phonon} = & -\frac{2\pi}{\hbar}\sum_{\vec{p}}\left|M_{x\ ph}(\vec{p}-\acute{k})\right|^2 \{[Q_k(1+q^{ph}_{k\ \acute{p}})(1+Q_{\acute{p}})-(1+Q_k)q^{ph}_{k\ \acute{p}}Q_{\acute{p}}] \\
& \times \delta(E_k - E_{\acute{p}} - \hbar\omega_{k\ \acute{p}}) + [Q_k q^{ph}_{\acute{p}\ k}(1+Q_{\vec{p}}) - (1+Q_k)(1+q^{ph}_{\acute{p}\ k})Q_{\vec{p}}] \\
& \times \delta(E_k - E_{\acute{p}} + \hbar\omega_{\acute{p}\ k})\} - Q_k / \tau_{opt}
\end{aligned}
\tag{5}
$$

Where excitons energy in $\vec{k}$ state is $E_{\vec{k}} = \hbar^2 k^2 / 2M_x$, excitons energy in $\vec{p}$ state is $E_{\vec{p}} = \hbar^2 p^2 / 2M_x$ and phonon energy is $\hbar\omega_{\vec{p}-\vec{k}} = \hbar v_s |\vec{p} - \vec{k}|$. The excitons mass is M_x. $Q_{\vec{k}}$ represents the exciton number in $\vec{k}$ state, $Q_{\vec{p}}$ represents the exciton number in $\vec{p}$ state and $n^{ph}_{\vec{p}-\vec{k}} = 1/[\exp(\hbar\omega_{\vec{p}-\vec{k}} / k_B T_b) - 1]$ represents the phonon occupation number. $M_{x-ph}\left(\vec{p} - \vec{k}\right)$ is known as the matrix element of the excitons phonon deformation potential interaction, and τ_{opt} is known as the radiative lifetime of excitons. $\left|M_{x\ ph}(\vec{p} - \vec{k})\right|^2 = \hbar D^2 |\vec{p} - \vec{k}| / (2V\rho v_s)$ is known as the excitons-phonon interaction term. Here D represents the deformation potential energy, V represents the crystal volume, ρ is known as the crystal density, the longitudinal acoustic sound velocity is v_s and the Dirac distribution is δ. The first term in the square brackets on the right-hand side of equation (3) is known as Stokes scattering and the second term is known as anti-Stokes scattering.

Excitons - excitons scattering term is given by [18, 19]

$$\left(\frac{dQ_{\vec{k}}}{dt}\right)_{polariton-polariton} d^3\vec{k} = \frac{2\pi}{\hbar} \cdot \frac{V^2}{(2\pi)^6} \int d^3\vec{k}\, d^3\vec{p}\, d^3\vec{p}_2\, d^3\vec{k}_2\, M_{matrix}^2$$

$$\times \{Q_{\vec{p}}Q_{\vec{p}_2}(1+Q_{\vec{k}})(1+Q_{\vec{k}_2})\}\delta(\vec{p}+\vec{p}_2-\vec{k}-\vec{k}_2) \qquad (6)$$

$$\times \delta(E_{\vec{p}}+E_{\vec{p}_2}-E_{\vec{k}}-E_{\vec{k}_2})$$

Here $E_{\vec{k}} = \hbar^2 k^2 / 2M_x$ is known as the excitons energy in $\vec{k}$ state. Similarly $E_{\vec{p}} = \hbar^2 p^2 / 2M_x$, $E_{\vec{p}_2} = \hbar^2 p_2^2 / 2M_x$ and $E_{\vec{k}_2} = \hbar^2 k_2^2 / 2M_x$ are known as the excitons energy in $\vec{p}$ state, $\vec{p}_2$ state, and $\vec{k}_2$ state, respectively. The excitons number, in $\vec{k}$ state is $Q_{\vec{k}}$, $\vec{p}$ state is $Q_{\vec{p}}$, $\vec{p}_2$ state is $Q_{\vec{p}_2}$, and $\vec{k}_2$ state is $Q_{\vec{k}_2}$. $M_{matrix} = \frac{4\pi\hbar^2 a_1}{M_x V}$ is known as the matrix element where a_1 is the scattering length. V is the volume. Here $e_{\vec{k}_2} = e_{\vec{p}} + e_{\vec{p}_2} - e_{\vec{k}}$ or $\vec{k}_2 = \sqrt{\vec{p}^2 + \vec{p}_2^2 - \vec{k}^2}$.

On the exciton occupation number, the complete action of Auger decay is given by [20] (Fig. **4.3**).

$$\left(\frac{\partial Q_{\vec{k}}}{\partial t}\right)_{Auger\ Decay} = -A_{pp}f(\vec{r})Q(\vec{r},\vec{k}) + \frac{1}{2}A_{pp}f(\vec{r})^2 \cdot \frac{(2\pi)^3}{\int d^3\vec{k}} \qquad (7)$$

The first term is for the two body decay and the second term is for recovery. The Auger constant is denoted by A_{pp}. The total exciton number along the direction of energy is the local density $f(\vec{r})$.

$$f(\vec{r}) = \frac{1}{(2\pi)^3} \int_k 4\pi k^2 Q(\vec{r},\vec{k})dk \qquad (8)$$

Auger constant $A_{pp} = 10^{18}$ cm^3 / ns has been applied in a recent experiment [10].

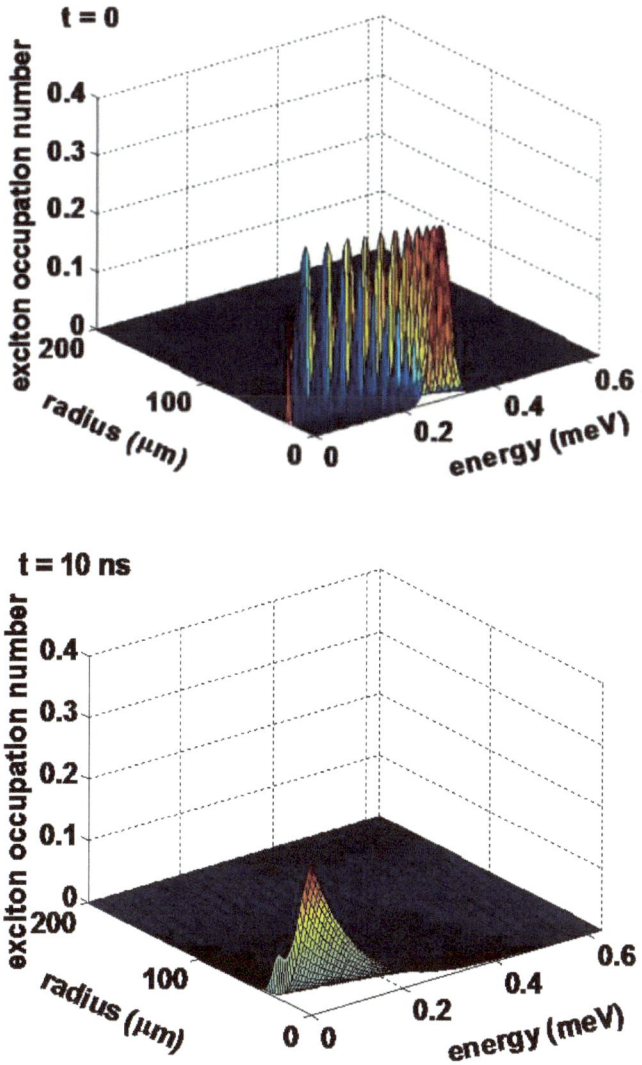

Fig. (4.3) The result of the numerical simulation with exciton-phonon and exciton-exciton scattering at temperature 0.1 K. The left hand side figure shows the initial distribution at t − 0 ns and the right hand side figure shows the result at 10 ns. 1.1×10^8 is the initial exciton number within the trap. Figure shows that at 10 ns, exciton distribution comes close to the zero energy of the trap.

But in the above case, the condensation threshold of the random phase approximation of the Boltzmann equation is no longer valid. Then quantum kinetics is the appropriate tool for that. A condensed excitons system is quasi equilibrium when the excitons' life time is longer than its relaxation time. In this case, the static properties of excitons can be described by the Gross-Pitaevskii equation. For a conserved number of particles, the time-independent Gross-Pitaevskii equation can be written as [21]:

$$\mu\Phi(r) = \left(-\frac{\hbar^2}{2m}\nabla^2 + V(r) + \rho\left|\nabla\Phi(r)\right|^2 \right)\Phi(r) \tag{9}$$

Here the wave function is Φ, the external potential is V, ρ is the coefficient related to the scattering length of the system's inter-particle interactions, μ is the chemical potential, $\hbar$ is the reduced Planck's constant and m is the mass of the boson.

The time-dependent Gross–Pitaevskii equation is [21]:

$$i\hbar\frac{\partial\Phi(r,t)}{\partial t} = \left(-\frac{\hbar^2}{2m}\nabla^2 + V(r) + \rho\left|\nabla\Phi(r,t)\right|^2 \right)\Phi(r,t) \tag{10}$$

The details about the Gross-Pitaevskii equation have been discussed in chapter 5.

RECENT DEVELOPMENTS IN THIS FIELD

Nowadays, the research interest in the field of the optical and optoelectronic properties of the two-dimensional group-VI transition metal dichalcogenide semiconductors, like MoS2, WSe2, *etc.* is expanding due to their importance in exciton physics and its application on devices [22]. The interlayer excitons between constituent monolayers are created due to type-II band alignment of the Van der Waals (vdW) heterostructures based on the transition metal dichalcogenides (TMDs).

TMD is an ideal metal platform for investigating the exciton transport phenomena. It is a semiconductor of type XY_2 where X is a transition metal atom such as W or Mo and Y is a chalcogen atom such as S, Se or Te. TMD has favourable electronic and mechanical properties. It shows a unique combination of atomic-scale thickness, direct band gap, and strong spin-orbit coupling. For these reasons, TMD is interesting for fundamental studies and for the applications in high-end electronics, optoelectronics, energy harvesting, flexible electronics, DNA sequencing and personalized medicine. Tightly bound excitons are formed by strong Coulomb interaction due to the reduced dimensionality and weak screening of TMD. These excitons are stable even at room temperature. TMD is a host of different types of excitons like bright, momentum-dark, and spindark excitons that control the spatial propagation. It produces a new type of spatially separated excitons with large in-plane or out-of-plane dipole moments that can be tuned *via* electric fields. For the improvement of excitonic integrated circuits, manipulation of the interlayer excitons in TMD vdW heterostructures has great potential [23]. In the integrated circuit, it bridges the optical communication and signal processing because it permits the photons and excitons to transform into each other.

From the ongoing research on the confined semiconductor excitons within an electrostatic trap of a GaAs bilayer heterostructure, it has been seen that during the cooling at sub Kelvin temperatures, optically bright excitonic states are heavily reduced [24]. On the other hand to keep the exciton population conserved, the optically dark and other accessible states become heavily increased. Therefore the combined spectroscopic signature contains mostly the dark Bose Einstein condensation of excitons. But in this case, the experiment is restricted for a critical temperature below 1K and dilutes the system within a limited number of densities.

A four component superfluid is produced by small number of optically bright excitons and optically dark excitons in the case of indirect excitons of GaAs quantum wells within electrostatic traps [25]. In this case, the macroscopic spatial coherence and quantized vortices are present below the critical temperature (about 1 K). But these behaviors have been seen for a small span of density in a dilute regime. In case of the spatially indirect excitons bound within a 10 μm electrostatic trap [26], the photoluminescence expands homogeneously. Below the critical density, the width of the spectral density decreases and a gray BEC occurs.

The researchers also studied the optical nature of quantum coherence in fully dark exciton condensates. It has been seen that the interaction between two fully dark exciton condensates generates bright interference fringes [27]. Due to the boson characteristics of excitons, the collision of two dark states produces bright states by the exchange of fermions.

In case of BEC of excitons in graphite, the BEC can be obtained at the critical temperature of 9.2 K and critical magnetic field of 47 Tesla [28]. Holes and electrons in Landau sub bands simultaneously cross the Fermi level and allow exciton formation at this critical field. By calculating the effective mass and the spatial splitting of the excitons in the basal plane, it has been seen that the degeneracy temperature of the excitonic fluid depends on the critical temperature. This evidence describes the reason of the field-induced transition in graphite that is not the general nature of three-dimensional electron systems.

Generally excitons condensation occurs at a very low temperature near absolute zero, but with smaller mass, excitons are expected to condense at higher temperature. In the case of interlayer excitons, those are produced electrically within MoSe2–WSe2 atomic double layers with a density of up to 1012 excitons per square centimeter [29]. It has been seen that the electroluminescence exhibits a threshold dependence on exciton density, a sensitive dependence on charge imbalance and critical fluctuations, those agree with exciton condensation. These observations remain up to about 100 K. These results open up more chances to explore exciton condensates at high-energy values, high temperature and exciton-controlled high-temperature superconductivity.

The researchers have also found the BEC at high temperature and they have realized that it is difficult to get high temperature excitonic BEC because of rigid restriction in symmetry, exciton binding energy, lifetime, and interaction [30] but it is possible. From the phase diagram of electron–hole systems, it can be seen that both BEC and super fluidity can be obtained at a high temperature and in a broad span of exciton density. This high-temperature behaviour of excitonic BEC is helpful for its experimental observation. Monolayer AuBr and BiS2 are recognized as good candidates for high-temperature excitonic BEC by applying the familiar approach to 2D materials in the database.

The study has also been performed to define excitonic properties in various types of excitonic insulator materials that can help make future electronic and optoelectronic devices. To do this, angle resolved photoemission spectroscopy has been used to generate the photoemission signal from spontaneously formed excitons within an excitonic insulator, Ta2NISe5 [31]. From the analysis of the direct excitonic photoemission spectra, the compact and highly anisotropic size of the excitons and their strong coupling are observed.

Large singlet exciton diffusion lengths are the main factors for high performance organic-based devices like photo detectors, photovoltaics and chemical sensors. Using ultrafast spectroscopic methods [32], investigation has been performed on the exciton dynamics of a two-dimensional covalent organic framework, COF-5. From the results, it has been seen that exciton diffusion in COF-5 is depend on the crystalline domain size. Research in this field will be helpful to make new and important optoelectronic devices.

Exciton transport and annihilation are two key procedures to control device efficiencies. A major challenge to know and understand the optoelectronic applications for low-dimensional nanostructures is to reach the long-range transport and suppressed annihilation while maintaining the large exciton binding energy. Two-dimensional hybrid organic-inorganic perovskites with strongly bound excitons and tunable structures are useful for optoelectronic applications [33].

In most of the classical organic semiconductor, the maximum size of the acceptor and donor domains depends on the exciton diffusion length which is used in organic photovoltaic cells. The efficiency of the solar cell is high if the materials are able to transport excitons over long distances. The exciton diffusion length can be measured using two different experimental procedures for a wide range of non-fullerene acceptor molecules [34]. These two different experimental techniques are formulated on ultrafast spectroscopy and photocurrent measurements. It is possible to understand and explain the exciton dynamics and extract the basic chemical design rules with the aid of quantum-chemical calculations from the results.

The hot exciton mechanism shows a new path for the progress of new-generation, high-efficiency, and pure-organic electroluminescent materials. The hot-exciton materials are actually organic light-emitting diode (OLED) materials that can control the non-radiative triplet excitons *via* reverse intersystem crossing (RISC) from high-lying triplet states to singlet states [35]. In this paper, the recent progress of the hot exciton process and basic principles about molecular design has been explained. Also hot exciton luminogens with their structure, property, interrelation and OLED applications have been discussed. It has been seen that it is low-cost, stable, pure-blue materials, which have the potential for industrial application. Therefore, the study on hot exciton materials is important for further progress of organic optoelectronic materials and devices.

The study on room temperature polariton BEC in organic perylenetetracarboxylic Diimide (PDI-O) single crystal micro ribbons in photonic circuits gives a new insight into making potential organic materials, which will be helpful for making polaritonic devices towards practical applications [36]. At room temperature, the excitons of PDI-O molecules interact strongly with microcavity photons due to Frenkel behaviour and large transition dipole moment. Huge coupling strength is generated in the microribbon cavities due to the high exciton densities in the single crystal, despite the low cavity Q. Huge coupling strength helps to make BEC of the polaritons.

A BEC of rubidium atoms has been produced in an earth orbiting research laboratory [37]. It has been observed that a BEC occurs within a weak trapping potential at sub-nano-kelvin temperature. In space, no microgravity is present in the environment. Therefore, researchers can use the full sensitivity of ultra-cold matter-waves to probe the basic physics and the principles of complex systems from which structure and dynamics come out. To do this experiment in an Earth-orbiting research lab, the multi-user research experimental set up is present in the International Space Station (ISS) since June 2018 after traveling from earth to space on board. It has been controlled by remote operation from the Jet Propulsion Laboratory.

APPLICATION OF BOSE EINSTEIN CONDENSATION OF EXCITONS

Bose Einstein Condensate can be used to make atomic laser, atomic clock, gravitational, rotational and magnetic sensors with excellent sensitivity.

Atom Laser

Atom laser is made from BEC of atoms. The output of the BEC of atoms, coupled using various techniques, is a coherent beam that acts as a wave called an atom laser. Some experimental research works on atom laser have been reported in references [38, 39]. The atom lasers are generated from an excited beam of atoms without making a trap (Fig. **4.4**).

The first atom laser was produced by a group at the Massachusetts Institute of Technology led by Wolfgang Ketterle in 1996. To do these experiments, researchers used an isotope of sodium and for output coupling technique, they used an oscillating magnetic field.

The physics of an atom laser is similar to that of an optical laser. The key distinction between them is that an optical laser is made by the light amplification by the stimulated emission of radiation whereas an atom laser is the output of the BEC of atoms. Atoms collide with themselves, and cannot be generated as photons can. Atoms have mass whereas photons do not have mass; therefore, atoms propagate at a speed below that of light.

First, an artificial operating atom laser was made by scientists Theodor Hänsch, Immanuel Bloch and Tilman Esslinger at the Max Planck Institute in Munich [41]. They have produced a well-controlled continuous beam ranging up to 100 ms. Before this, scientists produced only short pulses of atoms.

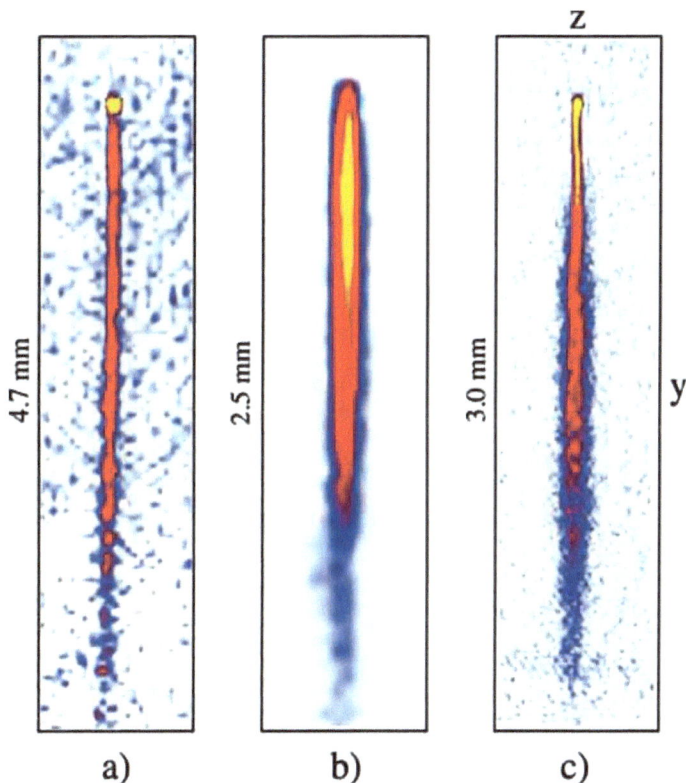

Fig. (4.4). Atom density distributions of atom lasers: **(a)** a well-collimated atom laser beam, **(b)** an ultra-high flux atom laser, and **(c)** atom beam combining atom laser and thermal emission. The density distributions were measured by absorption imaging after time-of-flight expansion. Reprinted (adapted) with permission from [40].

One use of an atom laser is atom holography. An atom laser can create holographic images with much higher resolution than conventional holography because the De Broglie wavelength of the atoms is much smaller than the wavelength of the light. Atom holography might be used to project a complex integrated-circuit of semiconductors which is just a few nanometers in length.

Another use of the atom laser is the atom interferometer. Atom interferometers are more sensitive than optical interferometers. This is because the atoms have mass, the de Broglie wavelength of the atoms is much smaller than the wavelength of light, and the internal structure of the atom is known. In an atom interferometer, an atomic wave packet is coherently split into two wave packets that go through different ways before reunion. Atom interferometers could be used to test quantum theory and to observe changes in space-time because it has high precision.

Atomic Clock

An atomic clock is one type of clock that calculates the time by monitoring the frequency of radiation of atoms. In this case, the normal oscillations of atoms act like the pendulum in an ordinary clock. Atomic clocks calculate the time more accurately than ordinary clocks because atomic oscillations have a much higher frequency and are much more stable. There are different types of atomic clocks, but they normally work on the same basic working principle, that is described below.

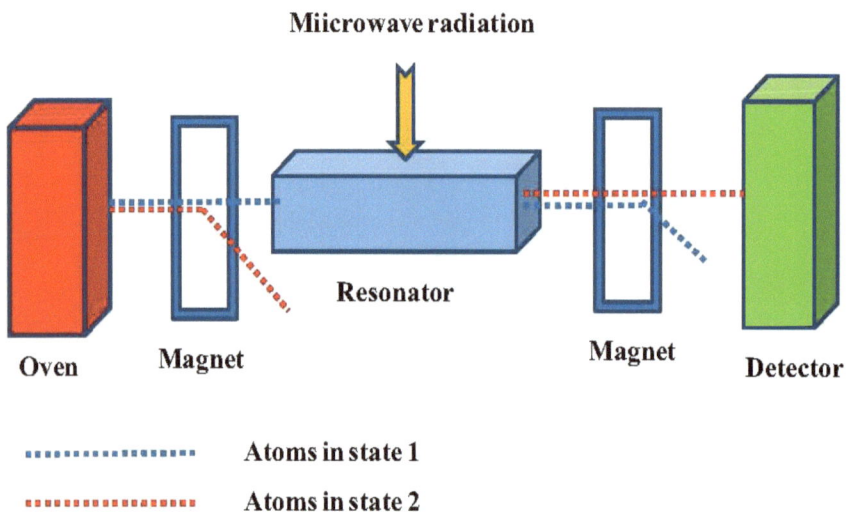

Fig. (4.5). Represents inner working of atomic clocks. It consists of one oven, two magnets, one resonator and one detector.

Working principle of atomic clocks: First the atoms are excited within an oven, then atoms create a beam. Now each atom has one of two probable energy states. These states are known as hyperfine levels. To understand the working principle, let us assume one is state 1 and another is state 2. An applied magnetic field extracts all atoms from state 2, therefore atoms remain only in state 1. The atoms of state 1 are passed through a resonator where microwave radiation is applied on these atoms. Meanwhile some of the atoms replace their state from 1 to 2. But the atoms in the back of the resonator are still in state 1.

They are eliminated by a second magnetic field. Now a detector counts all atoms that have replaced their state from 1 to 2 (Fig. **4.5**). The percentage of atoms that replace their state while going through the resonator depends on the frequency of the microwave radiation. The aim is to accurately tune the microwave frequency to the oscillation of the atoms, and then calculate it. After exactly 9,192,631,770 oscillations, a second has passed. The perfection of the atomic clock varies but it is always enhancing. The expected error of an atomic clock is only 1 second in about 100 million years. One of the world's most precise atomic clocks is the NIST-F1 in Boulder, Colorado.

Atomic clock is used to calculate the International Atomic Time (TAI), Coordinated Universal Time (UTC) and the local times over the world. Satellite navigation systems like GPS, GLONASS, and Galileo also depend on the accurate time calculation of atomic clock to estimate the positions correctly.

Gravitational, Rotational and Magnetic Sensors

Magnetic Sensors

A magnetic sensor detects the magnitude of magnetism and geomagnetism produced by a magnet or current. It is a sensor that converts the magnitude and variations of a magnetic field into electric signals. It is used to measure the change of properties like strength, direction, flux as well as disturbances within a magnetic field. There are several types of detection sensors that can work on some of the behaviours like light, pressure, and temperature. These detection sensors could be divided into two groups. One is used to measure vector components of the field, whereas another one is used to measure the total magnetic field (Fig. **4.6**).

Fig. (4.6). Magnetic Sensor.

The authors of a recent review article reviewed magnetic sensors and recent technologies depending on the magnetic sensors [42]. Sensing electric and magnetic fields with BEC has been studied in reference [43]. They claimed that using one dimensional BEC as a sensor it is possible to derive high resolution potential images along a line on a millimeter scale. Their experiments explain the unique sensitivity of the one dimensional BEC when applied to magnetic fields. They reported that this field sensor can be utilized to regenerate the current-density in a micro-fabricated wire. In future, this BEC sensor could be applied to obtain a deep knowledge of the local current flow for example in superconductors and two-dimensional electron gasses.

The magnetic sensor consists of a chip with a magneto resistive component which is used to detect a magnetic vector. A magnet used for magnetic vector biasing can be detected by the magneto resistive component. The chip is used for sensing the change within the magnetic vector. This vector measures the behavior of a magnetic body depending on the change of resistance value of the magneto resistive component. When the magnetic vector biasing takes place by the magnetic body, then there will be the motion inside the sensing chip. This sensor is used as a compass. The authors of reference [44] studied microscopic magnetic-field imaging.

Types of Magnetic Sensors

Magnetic sensors can be divided into three types based on detecting the dissimilarity of magnetic sensors. These three types are determined by how the sensor is applied with respect to the ever-present magnetic field of the earth. They are high sensitivity sensors, medium sensitivity sensors, and low sensitivity sensors (Fig. **4.7**).

<u>*High Sensitivity Sensors*</u>

These sensors are applied to measure extremely low values of the magnetic fields, approximately less than 10^{-5} G. The examples of low field sensors are superconducting quantum interference device (SQUID), nuclear procession and fiber optic. The applications of these sensors are mainly in nuclear as well as medical fields like brain function checking, magnetic abnormality observation, *etc.*

<u>*Medium Sensitivity Sensors*</u>

These sensors measure perturbation in the magnitudes and direction of Earth's field due to inducing or permanent dipoles. The range of magnetic field for this type of sensor ranges from 10^{-5} G to 1 G. It has many applications in the vehicle as well as navigation detection systems like magnetic compass, traffic control, *etc*. The example of medium sensitivity sensor is search coil magnetometer, fluxgate magnetometer and magneto resistive magnetometer.

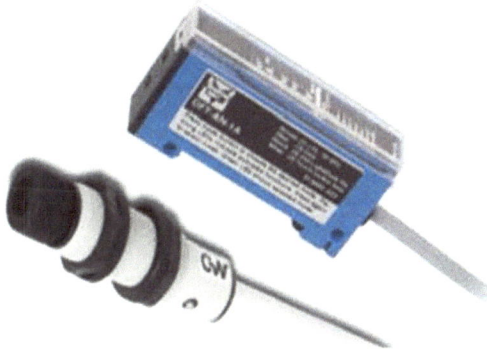

Fiber Optic Sensors

Hall Effect Magnetic
Sensor

Magnetoresistive Element
sensor

Fig. (4.7). Different types of magnetic sensors.

Low Sensitivity Sensors

These sensors are used to measure the magnetic fields stronger than earth's magnetic field approximately above 1 Gauss. Hall Effect sensor, search coil magnetometer, and magneto resistive magnetometer are the example of bias magnet field sensors. These sensors are used in noncontact switching, current measurement and magnetic memory readout.

Application of Magnetic Sensors

Small magnetic sensors are used to integrate in vehicles, mobile phones, medical devices, *etc.* for navigation, speed, position and angular sensing. In space sector applications in which mass, volume and power savings are very important issues, then these magnetic sensors are potential candidates. This work takes the magnetic technologies available in the marketplace and implements these in space applications.

Magnetic sensors are frequently applied for security and military purposes such as detection, discrimination and localization of ferromagnetic and conducting objects. These are also used for navigation, position tracking and anti-theft systems. The author of reference [45] studied the security applications of magnetic sensors.

Magnetic sensors have a lot of applications in medicine and biology. For example, magnetic tags can be attached to notice the presence of specific molecules. Magnetic micro beads were used as labels in a multi-analyst biosensor to detect DNA hybridization on a micro fabricated chip. The presence of the beads was detected by giant magneto resistance (GMR) magneto electronic sensors embedded in the chip [46]. The motion of the parts of the body such as the slight mechanical vibrations of eyelids or the articulator movements of the tongue during speech [47] can also be measured using magnetic sensors. The superconducting quantum interference devices (SQUIDS) have been used to measure extremely weak magnetic fields produced by the brain [48] and in other medical tests [49]. The authors of reference [50] studied magnetic sensors and their applications.

Gravity Sensor

Gravity sensors, also called gravimeters, are used to detect the minute localized changes in the force of Earth's gravity by variation in rock and other materials density in the underground. The sensitivity of the gravity sensor is capable of measuring the Earth tides. The Earth tides are like the ocean tides but these occur in the rock and soil and shift it to an almost unnoticeable degree. The Earth tides are produced when the planet's surface moves up and down. This occurs due to the changing distances between our planet, the sun and the moon. A decade ago, the gravity sensor was large, heavy and expensive. Therefore it had limited application in research and industry. But in 2016, by applying advanced technology it was possible to make micro sized gravimeters from silicon. It is a secure digital (SD) card size device that has a hollow center with 25 mg piece of silicon suspended by stiff fibrous structures about 5 μm thick. The silicon piece moves up and down slightly due to changes in the gravitational force. A light detector tracks the silicon piece's shadow and measures this movement.

Quantum Gravity Sensor

The latest development in gravity sensor is quantum gravity sensor. It is created by scientists Ernst Rasel of Leibniz University of Hannover and his colleagues [51]. This BEC gravimeter has a chip of a few centimeters in size which contains a tiny vacuum chamber. At the top of the chip, a laser and magnetic field trap are present that contains around 15,000 rubidium atoms. Within this trap, the atoms are cooled down to temperatures of a few nano-kelvin to create a BEC. Then the BEC is allowed to fall about 1 cm through the chamber and within this time, a series of laser pulses is fired on the BEC to measure the atoms into different paths to build an interferometer. It takes about 10 ms for the atoms to complete their free fall. The sensitivity could be increased if this time is expanded using a laser pulse to bounce the atoms back up when they reach the bottom of the chip. The atoms are then permitted to fall back down again and increase the free-fall time by a factor of five. The researchers [51] reported that their gravimeter worked in a rough environment without access to any vibration shielding, which caused much of the measurement in uncertainty. It is possible to improve the accuracy of this gravimeter by reducing the vibration and other improvements. The size of this gravimeter is also reasonable; it can fit inside of a backpack. Quantum sensing for gravity cartography has been studied by the authors of reference [52].

Application of Gravity Sensor

Gravity sensors have a lot of applications in research and industry. Micro-sized gravimeters could be used to sense the motion of magma in volcano zones. Then it is possible for researchers to understand the patterns of magma flow before eruptions. To search and rescue operations, drone-mounted gravimeters can be used to find underground cavities. Gravimeter is also used to map large-scale geological activities like change of the Earth's surface due to earthquakes or glaciers melting.

Quantum gravity sensors could be used to detect changes in microgravity using the principles of quantum physics. It is possible to know properties at the sub-molecular level. Therefore, it will improve mapping of geological activities below the ground level. It means it reduces the costs and delays for construction, rail, and road projects. It will improve the prediction of natural phenomena such as volcanic eruptions. It will help to discover the invisible natural assets, build information and understand the archaeological mysteries without damaging and digging.

Rotation Sensors

Rotation sensor is used to measure the rotary displacement in either clockwise or anticlockwise directions. There are different types of rotation sensor technologies; these are Hall Effect, Potentiometer, Encoder and Rotary Variable Differential Transformer (RVDT). A quantum device for detecting two-body interactions, scalar magnetic fields and rotations is suggested by using a Bose Einstein condensate (BEC) in a ring trap by the authors of reference [53].

Application of Rotation Sensor

Rotation sensors have a lot of applications in industries and environments. Some of these applications are: 1) Robotic applications – In this case, the sensors are used to measure the precise arm angles; 2) Medical applications – the rotary sensors are required in many medical devices for functioning properly; 3) CCTV cameras –to measure the precise angle of the camera; 4) Speed measurement – in various vehicles; 5) Motorsport applications – to measure the throttle position, steering and pedal position; 6) It is also used in Valve positioning, Crane positioning, Telescopic position and Machine tool.

CONCLUSION

There are several experimental procedures to observe the BEC of excitons. The Boltzman transport equation is useful to investigate the relaxation kinetics of excitons and the Gross-Pitaeveskii equation is used to investigate the BEC of excitons. Several research works have been performed in this field and some of them have been discussed here. The BEC of excitons has lot of applications. Atomic laser, atomic clock, gravitational, rotational and magnetic sensors are the few applications of it.

REFERENCES

[1]   Snoke, D.; Wolfe, J.P.; Mysyrowicz, A. Quantum saturation of a Bose gas: Excitons in Cu_2O. *Phys. Rev. Lett.,* **1987,** *59*(7), 827-830.
        http://dx.doi.org/10.1103/PhysRevLett.59.827 PMID: 10035881

[2]   Snoke, D.W.; Wolfe, J.P.; Mysyrowicz, A. Evidence for Bose-Einstein condensation of excitons in Cu_2O. *Phys. Rev. B Condens. Matter,* **1990,** *41*(16), 11171-11184.
        http://dx.doi.org/10.1103/PhysRevB.41.11171 PMID: 9993538

[3]   Lin, J.L.; Wolfe, J.P. Bose-Einstein condensation of paraexcitons in stressed Cu_2O. *Phys. Rev. Lett.,* **1993,** *71*(8), 1222-1225.
        http://dx.doi.org/10.1103/PhysRevLett.71.1222 PMID: 10055481

[4]   Proukakis, N.P. *Quantum gases: Finite temperature and non-equilibrium dynamics*; Cold Atoms SeriesImperial College Press: London, **2012,** 1, .

[5]   Snoke, D.; Littlewood, P. Polariton condensates. *Phys. Today,* **2010,** *63*(8), 42-47.
        http://dx.doi.org/10.1063/1.3480075

[6]   Fortin, E.; Fafard, S.; Mysyrowicz, A. Exciton transport in Cu_2O: Evidence for excitonic superfluidity? *Phys. Rev. Lett.,* **1993,** *70*(25), 3951-3954.
        http://dx.doi.org/10.1103/PhysRevLett.70.3951 PMID: 10054007

[7]   Mysyrowicz, A.; Fortin, E.; Benson, E.; Fafard, S.; Hanamura, E. Soliton propagation of excitonic packets and superfluidity in Cu_2O. *Solid State Commun.,* **1994,** *92*(12), 957-961.
        http://dx.doi.org/10.1016/0038-1098(94)90020-5

[8]   Benson, E.; Fortin, E.; Mysyrowicz, A. Study of anomalous excitonic transport in Cu_2O. *Phys. Status Solidi, B Basic Res.,* **1995,** *191*(2), 345-367.
        http://dx.doi.org/10.1002/pssb.2221910211

[9]   Mysyrowicz, A.; Benson, E.; Fortin, E. Directed beam of excitons produced by stimulated scattering. *Phys. Rev. Lett.,* **1996,** *77*(5), 896-899.
        http://dx.doi.org/10.1103/PhysRevLett.77.896 PMID: 10062934

[10]   Stolz, H.; Schwartz, R.; Kieseling, F.; Som, S.; Kaupsch, M.; Sobkowiak, S.; Semkat, D.; Naka, N.; Koch, T.; Fehske, H. Condensation of excitons in Cu_2O at ultracold temperatures: Experiment and theory. *New J. Phys.,* **2012,** *14*(10), 105007.
        http://dx.doi.org/10.1088/1367-2630/14/10/105007

[11]   Trauernicht, D.P.; Wolfe, J.P.; Mysyrowicz, A. Thermodynamics of strain-confined paraexcitons in Cu_2O. *Phys. Rev. B Condens. Matter,* **1986,** *34*(4), 2561-2575.
        http://dx.doi.org/10.1103/PhysRevB.34.2561 PMID: 9939950

[12]   Snoke, D.W.; Negoita, V. Pushing the Auger limit: Kinetics of excitons in traps in Cu_2O. *Phys. Rev. B Condens. Matter,* **2000,** *61*(4), 2904-2910.
        http://dx.doi.org/10.1103/PhysRevB.61.2904

[13]   Markiewicz, R.S.; Wolfe, J.P.; Jeffries, C.D. Strain-confined electron-hole liquid in germanium. *Phys. Rev., B, Solid State,* **1977,** *15*(4), 1988-2005.
        http://dx.doi.org/10.1103/PhysRevB.15.1988

[14]   Naka, N.; Nagasawa, N. Two-photon diagnostics of stress-induced exciton traps and loading of 1 s -yellow excitons in Cu_2O. *Phys. Rev. B Condens. Matter,* **2002**, *65*(7), 075209.
http://dx.doi.org/10.1103/PhysRevB.65.075209

[15]   Yoshioka, K.; Ideguchi, T.; Mysyrowicz, A.; Kuwata-Gonokami, M. Quantum inelastic collisions between paraexcitons in Cu_2O. *Phys. Rev. B Condens. Matter Mater. Phys.,* **2010**, *82*(4), 041201.
http://dx.doi.org/10.1103/PhysRevB.82.041201

[16]   Yoshioka, K.; Chae, E.; Kuwata-Gonokami, M. Transition to a Bose–Einstein condensate and relaxation explosion of excitons at sub-Kelvin temperatures. *Nat. Commun.,* **2011**, *2*(1), 328.
http://dx.doi.org/10.1038/ncomms1335

[17]   Som, S.; Kieseling, F.; Stolz, H. Numerical simulation of exciton dynamics in Cu_2O at ultra-low temperatures within a potential trap. *J. Phys. Condens. Matter,* **2012**, *24*(33), 335803.
http://dx.doi.org/10.1088/0953-8984/24/33/335803 PMID: 22836306

[18]   Som, S. Relaxation and condensation kinetics of trapped excitons at ultra-low temperatures: Numerical simulation. *Indian J. Phys. Proc. Indian Assoc. Cultiv. Sci.,* **2020**, *94*(10), 1603-1613.
http://dx.doi.org/10.1007/s12648-019-01592-7

[19]   Som, S. Homogeneous paraexciton dynamics at ultralow temperatures by numerical simulations. *J. Low Temp. Phys.,* **2019**, *197*(1-2), 44-60.
http://dx.doi.org/10.1007/s10909-019-02213-7

[20]   Hulin, D.; Mysyrowicz, A.; à la Guillaume, C.B. Evidence for bose-einstein statistics in an exciton gas. *Phys. Rev. Lett.,* **1980**, *45*(24), 1970-1973.
http://dx.doi.org/10.1103/PhysRevLett.45.1970

[21]   Voronych, O.; Buraczewski, A.; Matuszewski, M.; Stobińska, M. Numerical modeling of exciton–polariton Bose–Einstein condensate in a microcavity. *Comput. Phys. Commun.,* **2017**, *215*, 246-258.
http://dx.doi.org/10.1016/j.cpc.2017.02.021

[22]   Mueller, T.; Malic, E. Exciton physics and device application of two-dimensional transition metal dichalcogenide semiconductors *npj 2D Mater Appl.,,* **2018**, *2*, 29.
http://dx.doi.org/10.1038/s41699-018-0074-2

[23]   Jiang, Y.; Chen, S.; Zheng, W.; Zheng, B.; Pan, A. Interlayer exciton formation, relaxation, and transport in TMD van der Waals heterostructures. *Light Sci. Appl.,* **2021**, *10*(1), 72.
http://dx.doi.org/10.1038/s41377-021-00500-1 PMID: 33811214

[24]   Beian, M.; Alloing, M.; Anankine, R.; Cambril, E.; Gomez Carbonell, C.; Lemaître, A.; Dubin, F. Spectroscopic signatures for the dark Bose-Einstein condensation of spatially indirect excitons. *Europhys. Lett.,* **2017**, *119*(3), 37004.
http://dx.doi.org/10.1209/0295-5075/119/37004

[25]   Anankine, R.; Beian, M.; Dang, S.; Alloing, M.; Cambril, E.; Merghem, K.; Carbonell, C.G.; Lemaître, A.; Dubin, F. Quantized vortices and four-component superfluidity of semiconductor excitons. *Phys. Rev. Lett.,* **2017**, *118*(12), 127402.
http://dx.doi.org/10.1103/PhysRevLett.118.127402 PMID: 28388190

[26] Anankine, R.; Dang, S.; Beian, M.; Cambril, E.; Carbonell, C.G.; Lemaître, A.; Dubin, F. Temporal coherence of spatially indirect excitons across Bose–Einstein condensation: The role of free carriers. *New J. Phys.,* **2018**, *20*(7), 073049.
http://dx.doi.org/10.1088/1367-2630/aad30a

[27] Shiau, S.Y.; Combescot, M. Optical signature of quantum coherence in fully dark exciton condensates. *Phys. Rev. Lett.,* **2019**, *123*(9), 097401.
http://dx.doi.org/10.1103/PhysRevLett.123.097401 PMID: 31524492

[28] Wang, J.; Nie, P.; Li, X.; Zuo, H.; Fauqué, B.; Zhu, Z.; Behnia, K. Critical point for Bose–Einstein condensation of excitons in graphite. *Proc. Natl. Acad. Sci. USA,* **2020**, *117*(48), 30215-30219.
http://dx.doi.org/10.1073/pnas.2012811117 PMID: 33199600

[29] Wang, Z.; Rhodes, D.A.; Watanabe, K.; Taniguchi, T.; Hone, J.C.; Shan, J.; Mak, K.F. Evidence of high-temperature exciton condensation in two-dimensional atomic double layers. *Nature,* **2019**, *574*(7776), 76-80.
http://dx.doi.org/10.1038/s41586-019-1591-7 PMID: 31578483

[30] Wang, D.; Luo, N.; Duan, W.; Zou, X. High-temperature excitonic bose–einstein condensate in centrosymmetric two-dimensional semiconductors. *J. Phys. Chem. Lett.,* **2021**, *12*(23), 5479-5485.
http://dx.doi.org/10.1021/acs.jpclett.1c01370 PMID: 34086474

[31] Fukutani, K.; Stania, R.; Kwon, C. II *Detecting photoelectrons from spontaneously formed excitons;* Nat. Phys, **2021**, 17, pp. 1024-1030.
http://dx.doi.org/10.1038/s41567-021-01289-x

[32] Flanders, N. C.; Kirschner, M. S.; Kim, P. Large exciton diffusion coefficients in two-dimensional covalent organic frameworks with different domain sizes revealed by ultrafast exciton dynamics *J Am Chem Soc,* **2020**, *142*(35), 14957-14965.
http://dx.doi.org/10.1021/jacs.0c05404

[33] Deng, S.; Shi, E.; Yuan, L.; Jin, L.; Dou, L.; Huang, L. Long-range exciton transport and slow annihilation in two-dimensional hybrid perovskites. *Nat. Commun.,* **2020**, *11*(1), 664.
http://dx.doi.org/10.1038/s41467-020-14403-z PMID: 32005840

[34] Firdaus, Y.; Le Corre, V.M.; Karuthedath, S.; Liu, W.; Markina, A.; Huang, W.; Chattopadhyay, S.; Nahid, M.M.; Nugraha, M.I.; Lin, Y.; Seitkhan, A.; Basu, A.; Zhang, W.; McCulloch, I.; Ade, H.; Labram, J.; Laquai, F.; Andrienko, D.; Koster, L.J.A.; Anthopoulos, T.D. Long-range exciton diffusion in molecular non-fullerene acceptors. *Nat. Commun.,* **2020**, *11*(1), 5220.
http://dx.doi.org/10.1038/s41467-020-19029-9 PMID: 33060574

[35] Xu, Y.; Xu, P.; Hu, D.; Ma, Y. Recent progress in hot exciton materials for organic light-emitting diodes. *Chem. Soc. Rev.,* **2021**, *50*(2), 1030-1069.
http://dx.doi.org/10.1039/D0CS00391C PMID: 33231588

[36] Tang, J.; Zhang, J.; Lv, Y.; Wang, H.; Xu, F.F.; Zhang, C.; Sun, L.; Yao, J.; Zhao, Y.S. Room temperature exciton–polariton Bose–Einstein condensation in organic single-crystal microribbon cavities. *Nat. Commun.,* **2021**, *12*(1), 3265.
http://dx.doi.org/10.1038/s41467-021-23524-y PMID: 34075038

[37]   Aveline, D.C.; Williams, J.R.; Elliott, E.R.; Dutenhoffer, C.; Kellogg, J.R.; Kohel, J.M.; Lay, N.E.; Oudrhiri, K.; Shotwell, R.F.; Yu, N.; Thompson, R.J. Observation of Bose–Einstein condensates in an earth-orbiting research lab. *Nature,* **2020**, *582*(7811), 193-197.
        http://dx.doi.org/10.1038/s41586-020-2346-1 PMID: 32528092

[38]   Reinaudi, G.; Lahaye, T.; Couvert, A.; Wang, Z.; Guéry-Odelin, D. Evaporation of an atomic beam on a material surface. *Phys. Rev. A,* **2006**, *73*(3), 035402.
        http://dx.doi.org/10.1103/PhysRevA.73.035402

[39]   Kindt, L. Shock wave loading of a magnetic guide.*PhD thesis*; Utrecht University: Netherlands, **2011**. Available from: http://dspace.library.uu.nl/handle/1874/211584

[40]   Bolpasi, V.; Efremidis, N.K.; Morrissey, M.J.; Condylis, P.C.; Sahagun, D.; Baker, M.; von Klitzing, W. An ultra-bright atom laser. *New J. Phys.,* **2014**, *16*(3), 033036.
        http://dx.doi.org/10.1088/1367-2630/16/3/033036

[41]   Bloch, I.; Hänsch, T.W.; Esslinger, T. Atom Laser with a cw Output Coupler. *Phys. Rev. Lett.,* **1999**, *82*(15), 3008-3011.
        http://dx.doi.org/10.1103/PhysRevLett.82.3008

[42]   Khan, M.A.; Sun, J.; Li, B.; Przybysz, A.; Kosel, J. Magnetic sensors-A review and recent technologies. *Engineering Research Express,* **2021**, *3*(2), 022005.
        http://dx.doi.org/10.1088/2631-8695/ac0838

[43]   Wildermuth, S.; Hofferberth, S.; Lesanovsky, I.; Groth, S.; Krüger, P.; Schmiedmayer, J.; Bar-Joseph, I. Sensing electric and magnetic fields with Bose-Einstein condensates. *Appl. Phys. Lett.,* **2006**, *88*(26), 264103.
        http://dx.doi.org/10.1063/1.2216932

[44]   Wildermuth, S.; Hofferberth, S.; Lesanovsky, I.; Haller, E.; Andersson, L.M.; Groth, S.; Bar-Joseph, I.; Krüger, P.; Schmiedmayer, J. Microscopic magnetic-field imaging. *Nature,* **2005**, *435*(7041), 440.
        http://dx.doi.org/10.1038/435440a PMID: 15917796

[45]   Ripka, P. Security applications of magnetic sensors. *J. Phys. Conf. Ser.,* **2013**, *450*, 012001.
        http://dx.doi.org/10.1088/1742-6596/450/1/012001

[46]   Miller, M.M.; Sheehan, P.E.; Edelstein, R.L.; Tamanaha, C.R.; Zhong, L.; Bounnak, S.; Whitman, L.J.; Colton, R.J. A DNA array sensor utilizing magnetic microbeads and magnetoelectronic detection. *J. Magn. Magn. Mater.,* **2001**, *225*(1-2), 138-144.
        http://dx.doi.org/10.1016/S0304-8853(00)01242-7

[47]   Sonoda, Y. Applications of magnetometer sensors to observing bio-mechanical movements. *IEEE Trans. Magn.,* **1995**, *31*(2), 1283-1290.
        http://dx.doi.org/10.1109/20.364819

[48]   Masahiro, S.; Hiroaki, T.; Kunio, K.; Yasuhiro, H. MEG vision magnetoencephalograph system and its applications. *Yokogawa Tech. Rep.,* **2004**, *38*, 23-27. [English Edition].

[49]   Tsukamoto, A.; Saitoh, K.; Suzuki, D.; Sugita, N.; Seki, Y.; Kandori, A.; Tsukada, K.; Sugiura, Y.; Hamaoka, S.; Kuma, H.; Hamasaki, N.; Enpuku, K. Development of multisample biological immunoassay system using HTSSQUID and magnetic nanoparticles. *IEEE Trans. Appl. Supercond.,* **2005**, *15*(2), 656-659.
        http://dx.doi.org/10.1109/TASC.2005.849988

[50]   Wakai, R. Current and future technologies for biomagnetism *Proc. AIP Conf.,* **2004,** *724,* 14-19.

http://dx.doi.org/10.1063/1.1811813

[51]   Abend, S.; Gebbe, M.; Gersemann, M.; Ahlers, H.; Müntinga, H.; Giese, E.; Gaaloul, N.; Schubert, C.; Lämmerzahl, C.; Ertmer, W.; Schleich, W.P.; Rasel, E.M. Atom-chip fountain gravimeter. *Phys. Rev. Lett.,* **2016,** *117*(20), 203003.

http://dx.doi.org/10.1103/PhysRevLett.117.203003 PMID: 27886486

[52]   Stray, B.; Lamb, A.; Kaushik, A.; Vovrosh, J.; Rodgers, A.; Winch, J.; Hayati, F.; Boddice, D.; Stabrawa, A.; Niggebaum, A.; Langlois, M.; Lien, Y.H.; Lellouch, S.; Roshanmanesh, S.; Ridley, K.; de Villiers, G.; Brown, G.; Cross, T.; Tuckwell, G.; Faramarzi, A.; Metje, N.; Bongs, K.; Holynski, M. Quantum sensing for gravity cartography. *Nature,* **2022,** *602*(7898), 590-594.

http://dx.doi.org/10.1038/s41586-021-04315-3 PMID: 35197616

[53]   Pelegrí, G.; Mompart, J.; Ahufinger, V. Quantum sensing using imbalanced counter-rotating Bose–Einstein condensate modes. *New J. Phys.,* **2018,** *20*(10), 103001.

http://dx.doi.org/10.1088/1367-2630/aae107

CHAPTER 5

Bose Einstein Condensation of Polaritons

Abstract: Several experimental and theoretical procedures have been used to get the Bose Einstein Condensation (BEC) of polaritons. Nowadays, the investigation of the BEC of polaritons is vastly popular and increasing attention in atomic as well as solid-state physics because it has many technological applications. In this chapter, different experimental procedures, excitation, and observation methods are explained. Recent research developments in this field and the application of the BEC also are discussed.

Keywords: Application of polariton condensation, Bose Einstein Condensation of polaritons, Experimental techniques to get polariton condensation, Optical observation methods, Theoretical modeling to get polariton condensation.

EXPERIMENTAL TECHNIQUES TO GET BEC

The condensation of polariton can be obtained by reducing the temperature for the fixed particle number. Another way to get the polariton condensation is, for a particular temperature, the particle number has to be increased until the critical density for the condensation is reached. At a low concentration, excitons act as bosons but indeed excitons are made by fermionic particles. At a high concentration, excitons' fermionic behaviour comes into play. Therefore, it is difficult to get a proper density of polaritons, to obtain condensation. There are two ways to solve this problem. First, it is necessary to use the materials that have strong exciton binding energy. It will prevent the separation of the particles at high densities which is necessary for condensation. Another is to use the coupled quantum wells. In this case, it is possible to get the high exciton density before saturation effects occur because the coupling of quantum wells delocalizes the excitons.

The coupling of lower polaritons with external photons is the principle decay mechanism. This interaction can be observed directly. A photon is produced outside of the cavity when a polariton decays. By using standard optical techniques, this can be detected. The energy and in-plane wave vector should be conserved in this procedure. Other properties like angular momentum and phase coherence will also be conserved. It is possible to do the investigation on the emitted light from polaritons and also possible to know the system properties without any damage to the population. Polaritons can be created directly by pumping. In an alternate way, it is possible to stimulate electronic states at far higher energy than the polaritons.

These excited electronic states create a carrier of gas. If it cools down, then a quasi-thermal distribution of polaritons is formed.

Excitation Methods

To generate populations of polaritons, most of the experiments have used optical methods. This is the most convenient method. By using this process, it is possible to inject carriers electrically as shown in reference [1] where the researchers have work on the electronic polariton laser. To get the desired population, one must select the optical method in case of optical injection. There are two types of pumping, one is resonant pumping and another is non-resonant pumping. By resonant pumping, it is possible for experimentalists to create polariton populations directly and manipulate the quasicondensate properties. Non resonant pumping produces a quasi-thermal population of polaritons with the storage of free carriers and excitons.

Resonant Pumping

Resonant pumping produces a model of population or quasi-condensate with desired qualities. A condensate can be produced in a mono-energetic state by using a coherent source. Two condensates can interact by using two resonant beams [1]. To produce a condensate, it is also possible to apply a resonantly inserted pulse with low density [2, 3]. Due to the presence of a quasi-thermal background of polaritons, it would be possible to blend a thermally formed condensate. In this case, cool-down time is not required to form a condensate.

Non Resonant Pumping

In the case of non-resonant pumping at short wavelengths, the reflectivity minima of the Distributed Bragg reflector (DBR) are used. Then it is possible to create free carriers in the quantum wells with good productivity.

To know the full picture of polaritons properties, non-resonant pumping is often the optimal solution. In case of low polariton concentration, discharge may be seen only from the lower states. The excitons in the quantum wells scatter many times to come into the polariton states. For this reason, it can be assumed that most of the properties of the pump like polarization and phase coherence will be lost before becoming polariton. This behavior is most favorable for the studies of condensation. Several types of excitation lasers are used for excitation, depending on the type of experiment. These are described below.

Continuous Wave (CW) Multimode Titanium Sapphire

This is a standard Z cavity multimode Ti:Sapphire laser with a triple birefringent plate filter that allows a multimode emission of a relatively large width nearly 10 GHz. The Ti:Sapphire crystal has a typical emission spectrum starting from nearly 690 nm to 1 micron. The cavity length is on the order of 60 cm according to the mode spacing that is on the range of 250 MHz. The presence of multiple modes produces noisy lasing that gives a clear mark in the polariton luminescence. The noise is adequately slow to follow the excitonic reservoir and therefore, there is a time-dependent blueshift of the polariton condensates *via* the exciton polariton collisions that enlarge the spectra notably (roughly by a factor of 5). It has been proved that a stable lasing of this laser can be obtained at 695 nm, because this wavelength is at the border region of the gain spectrum. The output power of this laser is nearly 600 mW for 5.5 W of pump power.

Continuous Wave (CW) Multimode Laser Diodes

The continuous wave multimode laser diodes are used to overcome the noisy multimode CW laser in the experiments. These diodes are not tunable. They were lasing at 685 nm in the vicinity of the second reflectivity minimum after the stop band of the sample.

The main disadvantages are the low force, the use of two diodes in a perpendicular polarization and orientation in order to couple them in the same excitation path without losing information. The multimode emission had a mode spacing of the order of nearly 25 GHz with extremely small cavity length. The fluctuations of resultant intensity have much rapid time span than that for controlling the excitonic reservoir. In contrast, when mono-mode laser is excited with the diodes, it has been seen that a high number of condensate modes is visible on average. This gives the conclusion that the rapid time span noise of the diodes produces additional excitations to the condensate.

Continuous Wave (CW) Monomode Titanium Sapphire

The way of resolving all the excitation noise induced problems is the use of a true monomode laser. The monomode laser is a coherent model MBR-EL that is a bow-tie ring cavity. It contains a thin etalon and electronic stabilization for mode-hoping free single mode operation. To get unidirectional lasing, an isolator is added in the cavity that forces the photons to circulate only in one direction. Coarse tuning is done by a birefringent filter and fine tuning is implemented by tilting the thin etalon.

The standard output power is on the order of 1 Watt with a 10% efficiency that is unusual for the laser wavelength, around 695 nm in the border of the gain curve.

Mode Locked Pulsed Titanium Sapphire

The investigation of the dynamics of polariton condensates is an attractive topic as it is the only way to examine the time-dependent phenomena. For this reason, it is required to use femto second pulsed lasers for the excitation. To get pulsed lasing, mode locking is performed by an acousto-optic modulator that spectrally overlaps in subsequent modes. This laser can be arranged in a straight line for lasing at 695 nm that gives Gaussian temporal pulses of average width 250 femto second. The efficiency of this laser is nearly 5% for this particular wavelength.

Inhomogeneous Optical Pumping

Two common inhomogeneous optical pumping methods for generating and replenishing the polariton condensates are Gaussian pump and Laguerre-Gaussian pump, respectively.

Gaussian Pump

A Gaussian beam distribution at position z = 0 can be written as:

$$f(x,y) = A \exp\left[-\left\{ \frac{(x-x_0)^2}{2\omega_x^2} + \frac{(y-y_0)^2}{2\omega_y^2} \right\} \right] \tag{1}$$

Here the amplitude is A, the center of the pump spot is $(x-x_0)^2$, ω_x and ω_y control the width of the beam along the x and y direction, respectively. The intensity of the beam is $I = |f|^2$. In the experimental cases, the pump beam may be not perpendicular to the sample. Using a cylindrical lens, the elongated Gaussian distribution can be shaped [4]. The Intensity distribution of a Gaussian beam is shown in Fig. 5.1. If we put the value of A, x_0, y_0, ω_x and ω_y in equation (1) and calculate it, we will get the value of f(x, y) in terms of x and y. Then we can calculate the intensity using equation $I = |f(x,y)|^2$. Now if we plot the intensity as a function of x and y, we will get Fig. 5.1. In case of (Fig. **5.1, a**) the values are

$x_0 = 0$, $y_0 = 0$, and $\omega_x = \omega_y = 30\,\mu m$ and in case of (Fig. **5.1, b**) the values are $x_0 = 0$, $y_0 = 0$, $\omega_x = 30$ and $\omega_y = 20\,\mu m$.

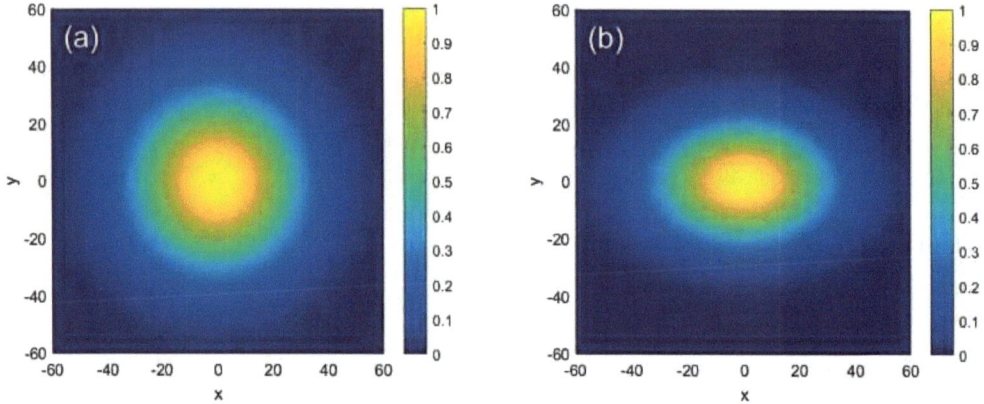

Fig. (5.1). Intensity distribution of a Gaussian pump given by Eq. (1) (a) $\omega_x = \omega_y = 30\ \mu m$ (b) $\omega_x = 30\ \mu m$ and $\omega_y = 20\ \mu m$.

Laguerre-Gaussian Pump

The Laguerre-Gaussian (LG) beam is obtained from higher-order mode solutions of the paraxial Helmholtz equation with circular symmetry. It can be written as [5],

$$f(r,\phi,z) = \frac{A}{\sigma(z)}\left[\frac{r\sqrt{2}}{\sigma(z)}\right]^{|l|} \exp\left[-\frac{r^2}{\sigma(z)^2}\right] L_p^{|l|}\left[\frac{2r^2}{\sigma(z)^2}\right]$$
$$\times \exp\left[-ik\frac{r^2}{2R(z)}\right]\exp(il\phi)\exp(-ikz)\exp[i\psi(z)]$$

(2)

Where, the cylindrical coordinate variables are r, ϕ and z, normalization constant is A, the Gouy phase, $\psi(z) = (|l| + 2p)\arctan(z/z_R)$ is a phase shift gradually acquired by a beam around the focal region, L_p^l is the generalized Laguerre polynomials, and l is the azimuthal mode number. R(z), evolving beam width, is the reciprocal of the curvature. Along the beam at position z from the focus, the spot size parameter R(z) can be written as:

$$R(z) = z\left[1 + \left(\frac{z_R}{z}\right)^2\right] \tag{3}$$

The curvature of the wave fronts $\sigma(z)$ can be written as:

$$\sigma(z) = \sigma_0\sqrt{1 + \left(\frac{z}{z_R}\right)^2} \quad \text{where} \quad z_R = \frac{\pi\sigma_0^2 n}{\lambda} \tag{4}$$

z_R is called the <u>Rayleigh</u> length and n is the refractive index of the medium. When the beam profile is forwarded (or backwarded) by l, it completes a 2π phase rotation. This is an example of an optical vortex of topological charge l [6], and can be related with the orbital angular momentum of light in that mode. The positive integer parameter p controls p + 1 intensity dips (nodes) in the radial direction.

The orbital angular momentum carried by the LG beam can be directly transferred to a polariton BEC under the coherent pumping scheme [7]. Under the incoherent pumping scheme, the phase information of the pump will be washed out so that only the intensity of the beam would be a concern.

General Excitation and Detection Setup

A general time-resolved photoluminescence excitation and detection setup in back-reflection geometry is shown in the figure below (Fig. **5.2**). This optical setup provides high flexibility for choosing the different excitation sources and the various detection setups. The main excitation sources that can be applied are CW or pulse lasers. Continuous wave excitation was executed in a quasi CW manner in order to keep away heating of the sample. The wave excitation was performed by modulating the laser intensity.

The polariton density is the important parameter that was applied to direct the polaritons through the phase transition of the condensation. This density is directly connected to the intensity of the excitation laser that was controlled by a thick (1 cm) variable rotational neutral density filter. The laser beam was focused onto the surface of the sample from a long working distance through a high numerical aperture microscope. The advantage of this type of microscope is that it can provide very high real space resolution. To increase the size of the excitation spot in real space, a lens of 250 mm is used to focus the beam on the back focal plane of the microscope. To provide the required size, we decrease the size of the excitation spot in Fourier space, having an opposite effect in real space. It also reduces the angular spread of the excitation mostly around the zero value of the momentum vector, $k_{//}$ = 0. To get the desired output, a lens is used to form a real space image of the sample surface by using a CCD camera or the streak camera. In the output, both luminescence and reflected excitation lasers are present. Therefore the proper way of removing the reflected laser light is the use of a spectral filter.

Flexible rotational neutral density filters that can be modified easily, are used in front of all the detection devices to prevent saturation. All spectral studies are conducted by a high-resolution spectrometer. Till now most of the resolved studies are performed by pulsed laser excitation and a streak camera on the detection side.

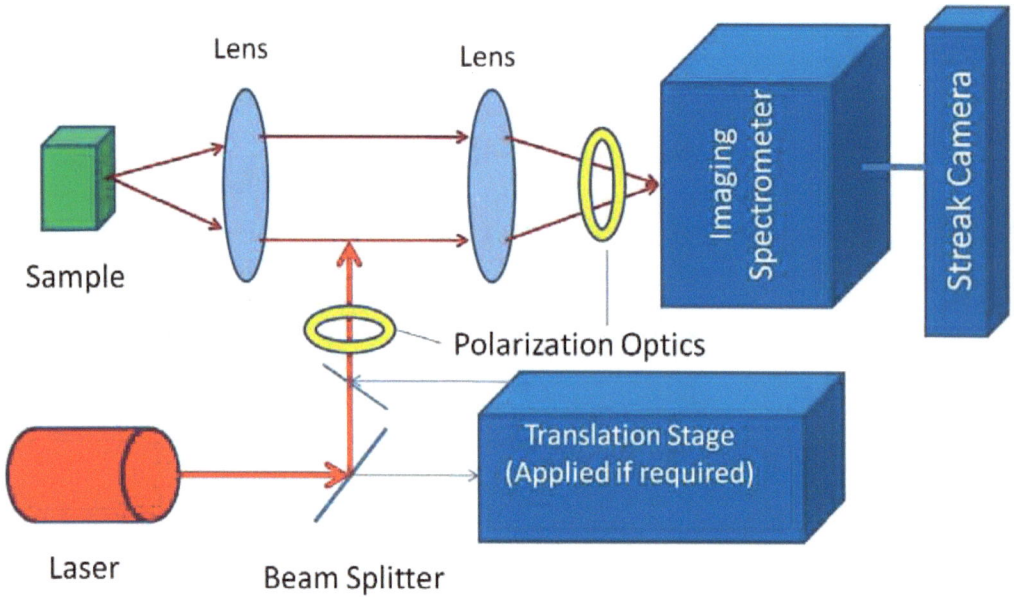

Fig. (5.2). Experimental set up for excitation and detection of polaritons in semiconductor nanostructures. First, the laser beam comes from the source and by using a beam splitter, it is directed towards the surface of the sample. To make a pulsed laser and therefore to get another laser pulse translation stage is used. Another beam splitter is used to direct the beam towards the surface of the sample. To increase the size of the excitation spot in real space, a lens is used in front of the sample. Another lens is used to focus the beam on the spectrometer and then the image is taken by using a streak camera. When the laser beam goes toward the surface of the sample, just before the surface of the sample, polarization optics is used and after excitation when the light beam goes to the spectrometer, polarization optics is used. These specific polarizations are used to prevent unwanted back reflections from optical components.

Optical Observation Methods

In the decay process, energy, momentum and spin angular momentum must be conserved. When a polariton decays, it is possible to investigate the emitted photons and from the emitted photons, we are able to get the information about the polaritons. Then it is possible to prove that in-plane momentum should be conserved. Fig. (**5.3**) represents the in-plane momentum conservation across optical interfaces and a comparison of the wave vector of the polariton in the cavity and outside of the sample. Here, a wave vector decomposes into its in-plane and out of plane components. The in-plane component is k_x and the out of plane component is k_z. The middle layers of a Distributed Bragg Reflector (DBR) structure have no effect on these conservation laws.

It is possible to calculate the interrelation between the wave vectors of light in the cavity and in air by using Snell's law. If the wave vector in the air is $\bar{k}$ and the wave vector inside the medium is $\bar{k}'$. Then the relation between the magnitudes of wave vector, by using the conservation of energy, is

$$E = \frac{\hbar c}{n_c}|k'| = \hbar c |k|$$

$$\text{Or } |k'| = n_c |k|$$

$$\text{or, } n_c = \frac{|k'|}{|k|} \tag{5}$$

As we know that the refractive index of air is nearly equal to one, here refractive index of air is taken as one. Now calling θ (θ') to be the angle of $\bar{k}(\bar{k}')$ and applying it to equation (5) above, we get,

$$n_c \sin(\theta') = \sin(\theta) \qquad \text{as } n_c \sim 1 \text{ in air} \tag{6}$$

Now putting the value of n_c from equation (5), we get,

$$|k'|\sin(\theta') = |k|\sin(\theta)$$

(7)

If we see the triangle on the upper side of (Fig. **5.3**), which is in the air, we get,

$$|k|\sin(\theta) = |k_x|$$

(8)

And from the triangle on the lower side of (Fig. **5.3**), which is in the cavity, we get,

$$|k'|\sin(\theta') = |k_x'|$$

(9)

Based on equation (7), (8) and (9), we can say

$$|k_x'| = |k_x|$$

(10)

Therefore, the in-plane momentum is conserved.

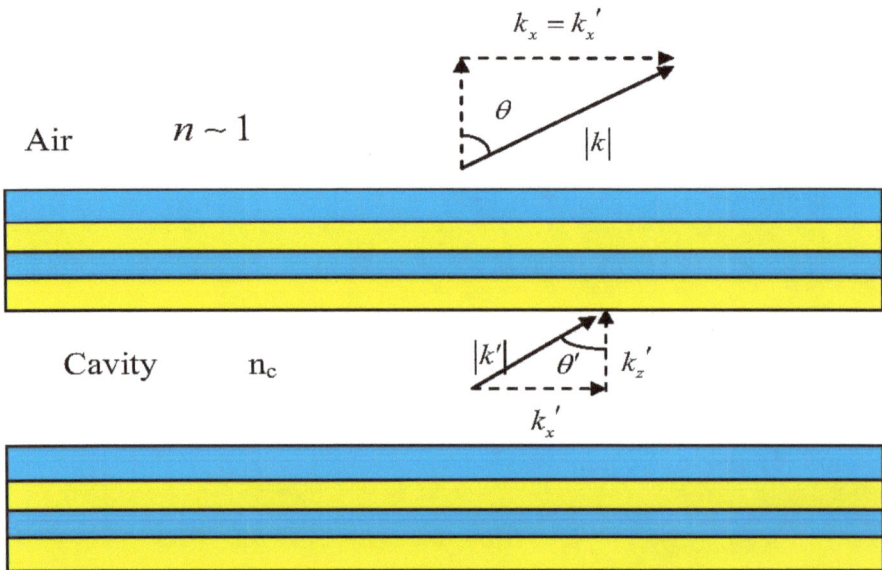

Fig. (5.3). In-plane momentum conservation across optical interfaces. Comparisons between the wave vector of the polaritons inside the cavity and outside the sample.

This conservation of in-plane momentum also indicates that the direction of $\vec{k}'_x$ is conserved, because the refracted ray stays in the same plane of incidence.

Real Space Imaging

Position-space or real-space imaging is the most straight forward to study polaritons where researchers can apply usual optics to accumulate and take the image of the light released from the sample. For imaging and measuring light, all standard optical techniques are important. We can use wave plates, polarizer, wavelength dependent filters, CCD cameras, *etc.* to manipulate and detect the light.

A typical imaging setup contains an infinity corrected microscope and an imaging lens. In case of collected light, infinity corrected lenses give the minimum spherical aberrations. It is possible to get a moderate to high magnification of an object by using a microscope. To take an image of a certain state, we can use Fourier filtering because in-plane momentum upon emission is conserved. As BEC can exist only in the ground state, so in order to get the image of a condensed portion of a gas, only $k_{\text{II}} \approx 0$ is required. To extract the intense reflected laser light from the excitation, this process can be useful. By setting an iris or pinhole at the Fourier image of the sample, this filtering is achieved. For imaging setup, this process is useful to trace the rays. The final magnification can be selected by simply changing the final imaging lens which is a quality of this optical design. But one of its difficulties is that if the final image is generated on a spectrometer slit, then such an effect needs an adjustment of the total path length from the sample to the final image. Another point is that while applying a microscope objective, it will be unsuitable to use Fourier filters at the first momentum space image because this would also obstruct other light through the objective.

Within the large compound lens, the Fourier image may be formed. Spectrometers have built-in numerical aperture constraint. In case of enlarged source over the full numerical aperture, some of the light can be cut as it passes through the spectrometer to get good image. Though, a good image is generated on the spectrometer entrance slit for the medium size source. It is also possible to take an image of the different parts of the field of view. We can get both a real-space and momentum-space filtering by extending the imaging system. Due to this reason, the undesirable scattered or reflected light can be removed from the final image. But in this case, the filtering does not interfere with the excitation because the images are far from the injection optics.

To select the field of view, the inclusion of the final imaging lens adds flexibility. This final lens can be inserted on a micrometer driven translation stage that allows the observer to take image at different regions of the field of view. Due to the combined nature of the microscope, objective-imaging lens system, translating only one of these can easily add aberrations into the image. Furthermore, depending on the experimental processes, the movement of the objective can change the pump conditions. In Fig. (**5.4**), the experimental set up for real space imaging is shown.

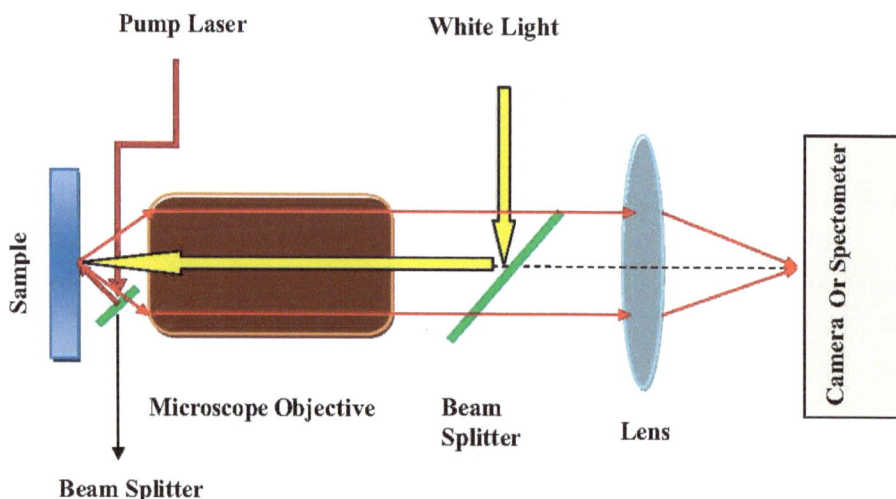

Fig. (5.4). This figure represents the experimental set-up for real space imaging. For the spectral reflectivity measurement, white light (yellow arrow) has been inserted into the microscope objective using a beam splitter. The pump laser is inserted into the sample using a beam splitter. The beam coming out from the sample is focused and managed by using a separate lens. The collected light may be luminescence or scattered light. This collected light is imaged directly by the camera or further processed before collection.

Momentum Space Imaging

The in-plane momentum can be measured by the angle-resolved imaging. As we know that the polaritons show an interrelationship between the in-plane wave vectors inside the exterior photons, therefore, it is important to measure the in-plane momentum. This type of imaging is known as momentum space imaging. In this case, a lens is used to do Fourier transform of light which is released from the sample. In the geometrical optics for an ideal lens, the positive focal plane is a proper angle-transformed image position of light passing through it.

These real and momentum images allow us to extract the undesired things from the data. It is helpful to take a final image for both real and momentum space after extraction of the undesired things. The real and angle-resolved images should be projected onto the same plane because it is tough to shift the detection apparatus like cameras, spectrometers or streak cameras, *etc.* Therefore in this system, the final real-space imaging lens is generally put back with a single lens or lens system to generate the angle-resolved image on the sample of the detector plane.

The calibration of momentum space images is complicated than that of real-space. While the momentum space images can be measured by applying a test sample with known properties or moving the sample with a known distance (usually using a micrometer). The simple way to calculate this magnification is to apply a calibrated iris between the microscope objective and the sample. This creates a shadow in the k-space that depends on the diameter of the iris and the distance between the objective and the sample. It requires a source of light that can either emitted from a point in the sample plane or focused through the objective. Generally, a flexible iris is used and calculates a series of iris diameters in the aperture. Here the problem is, each time it needs to remove the iris that can change the position relative to the optical axis. Errors can come from the spatial size of the light source and from the incorrect position of the iris relative to the optical axis. In case the working distance of the objective is too short to permit inserting an iris, then this method will not work. It is important that by using this method, the emission angle from the sample is determined.

As we are concerned about the momentum vector $k_\parallel$ and since $k_\parallel = |k| \sin(\theta) \infty E \sin(\theta)$, we can only appropriately calibrate the k-axis for energy related data. Generally, k-axis is approximated as uniform in a given image because the respective change in $|k|$ is slightly over the span of the data accumulated from the polaritons. It is also important to know that in the spectrally resolved case, the k-resolved data is more meaningful.

Spectroscopy

In case of optical experiments, the spectrometer is an important instrument to spectrally disperse the light. Generally, a grating is used to spread the light into its composing colours. This allows us to measure the wavelength of light in an image.

According to the working principle of grating, in order to gain the spectral dimension, it requires taking only one dimension of an image. Therefore, we need to choose a portion of an image like x vs. y, then this is dispersed in energy for example energy vs. x. To get the spectral configuration of a complete x-y image, one must take a series of spectral pictures while the x-y image goes through the spectral slit.

To take a one dimensional image, it is not difficult, but one should be careful when the system is not isotropic. Often the real and k-space data have destroyed the symmetries. In many cases, it is difficult to correctly place the gradient or a structure on the sample with the spectrometer slit.

The limiting numerical aperture is another important point of spectrometers. Spectrometers have long optical path lengths because the resolving power of the spectrometer increases with the rising length. This means that light can scatter much more within the spectrometer.

Time Resolved Imaging

Generally, the polaritons stay in stable position rather than the exact thermal equilibrium state because polaritons decay continuously. It is possible to work in the steady state regime depending on the experiment where the pumping is done by the CW laser. Even in many systems, the polariton's decay may be faster than the filling rate from the excited states. Therefore, it is possible to get a steady state population of polaritons from the gain of the CW laser into higher energy states. This is known as quasi-CW regime. To get steady state population for longer than the transient state population, a chopped CW laser pumping is being used. This is used to reduce the heating of the sample, as well as retain the population in the CW state. It is possible to characterize the gradual process of change and development of the constituent polaritons if we can calculate the polaritons on an appropriate timescale. In the observable range, polaritons have lifetimes in the order of 1 to 100 picoseconds (ps) and go through the in-plane on the order of 4% of the speed of the light. Two particular methods can be used to probe the system on the picoseconds timescale. One can use the streak cameras which transform the temporal dispersion into spatial displacement. This method is applicable when the gradual progress of a population follows a femto-second or picoseconds excitation. When two pulses of lights are impinged on a sample with properly organized time delay, then a pump-probe technique can be used. Generally, the delay between the pulses changes the path length of one arm before entering but that can be controlled with a delay micrometer.

Theoretical Modeling to get BEC

Polaritons condensation kinetics involves creation, relaxation and decay processes since polariton has a lifetime of 1 to 100 ps. The semi classical Boltzmann rate equation can be used to describe the relaxation kinetics of polaritons [8]. This is discussed in chapter 4. The Gross-Pitaevskii equation can be used to describe the ground state dynamics of polaritons. The name Gross-Pitaevskii equation came from the name of Scientist Eugene Paul Gross and Lev Petrovich Pitaevskii. The Gross-Pitaevskii equation describes the ground state of the quantum system of ideal bosons using the pseudo potential interaction model and the Hartree–Fock approximation.

When a large fraction of bosons occupy the lowest quantum state, it is known as Bose Einstein condensate and these bosons can be described by the same wave function. A free quantum particle can be described by a single-particle Schrödinger equation. By using appropriate many-body Schrödinger equation, it is possible to take into account the interaction between particles.

It is possible to approximate the true interaction potential when the mean distance between the particles in a gas is larger than the scattering length. At sufficiently low temperature, the scattering process can be estimated by the s-wave scattering when the de Broglie wavelength is much larger than the span of boson–boson interaction. In that case, according to the pseudo potential model, Hamiltonian of the system of n bosons can be written as:

$$H = \sum_{j=1}^{n}\left(-\frac{\hbar^2}{2m}\frac{\partial}{\partial r_j^2}+V(r_j)\right)+\sum_{j<k}\frac{4\pi\hbar^2\sigma_s}{m}\delta(r_j-r_k) \qquad (9)$$

Here the external potential is V, the mass of the boson is m, $\hbar$ is the reduced Planck's constant, the boson-boson scattering length is σ_s and the Dirac delta-function is $\delta(r_j-r_k)$.

According to the Hartree–Fock approximation, we can write the total wave function of the system Φ as a product of single wave function φ in the following way.

$$\Phi(r_1,r_2,r_3,......r_n) = \varphi(r_1)\varphi(r_2)......\varphi(r_n) \qquad (10)$$

Where r_j is the coordinate of the jth boson and j=1, 2 ...n.

According to the variation method, if the single-particle wave function satisfies the following Gross–Pitaevskii equation,

$$\left(-\frac{\hbar^2}{2m}\frac{\partial}{\partial r^2} + V(r) + \frac{4\pi\hbar^2\sigma_s}{m}|\varphi(r)|^2 \right)\varphi(r) = \mu\varphi(r) \qquad (11)$$

Here μ is the chemical potential. The total wave-function minimizes the expectation value of the model Hamiltonian under normalization condition $\int dV|\varphi|^2 = n$. Therefore the ground state of the system can be described by such a single-particle wave function.

Gross-Pitaevskii equation, sometimes referred to as the nonlinear Schrödinger equation, is alike in form with the Ginzburg–Landau equation. The origin of the non-linearity of the Gross-Pitaevskii equation lies in the collision between the particles. In the Gross-Pitaevskii equation, if we put the coupling constant of the collision is zero, then the single-particle Schrödinger equation comes up that describes a particle within a trapping potential. The Gross-Pitaevskii equation has some limitations in the weakly interacting region. Its form is like Schrödinger equation but it has one additional interaction term. The s-wave scattering length σ_s is proportional to the coupling constant ρ of two interacting bosons.

$$\rho = \frac{4\pi\hbar^2\sigma_s}{m} \qquad (12)$$

Here, $\hbar$ is the reduced Planck's constant, and m is the mass of the boson. The energy density is:

$$\xi = \frac{\hbar^2}{2m}|\nabla\Phi(r)|^2 + V(r)|\Phi(r)|^2 + \frac{1}{2}\rho|\Phi(r)|^4 \qquad (13)$$

Where the wave function is Φ, and the external potential is V. For a conserved number of particles, the time-independent Gross–Pitaevskii equation is:

$$\mu\Phi(r) = \left(-\frac{\hbar^2}{2m}\nabla^2 + V(r) + \rho|\nabla\Phi(r)|^2 \right)\Phi(r) \tag{14}$$

Here μ is the chemical potential and the number of particles is related to the wave function by:

$$n = \int |\Phi(r)|^2 \, d^3r \tag{15}$$

It is possible to obtain the structure of a Bose Einstein condensate with different external potentials by using the time-independent Gross–Pitaevskii equation. From the time-dependent Gross–Pitaevskii equation, the dynamics of the Bose Einstein condensate can be investigated. Therefore it is possible to find the collective modes of a trapped gas. The time-dependent Gross-Pitaevskii equation is:

$$i\hbar\frac{\partial\Phi(r,t)}{\partial t} = \left(-\frac{\hbar^2}{2m}\nabla^2 + V(r) + \rho|\nabla\Phi(r,t)|^2 \right)\Phi(r,t) \tag{16}$$

Here, t represents the time.

Recent Developments in this Field

Nowadays, research in this field is very interesting and a growing field. It has lots of technological applications. Several theoretical and experimental works have been done in this field [9-11].

The study of polariton dynamics in the strong coupling regime is one area within this field. When photoactive molecules collide with confined light mode electromagnetic field then the system goes into the strong coupling regime and polaritons are formed. The relaxation process strongly depends on the Stokes shift and therefore, polariton dynamics can be controlled by the proper choice of molecules. That is important for practical applications like polariton lasing. Therefore, the stokes shift is very important to analyze the experiments on strongly coupled molecular cavity systems [12]. Stokes shift is the energy, wave number or frequency difference between the maxima of the absorption and the emission spectra of the same electronic transition. The name stokes shift came from the name of Irish physicist George Gabriel Stokes.

The Rabi splitting in the dispersion has shown the ability to control the chemistry of the coupled molecules. Using two-dimensional infrared and filtered pump–probe spectroscopy, researchers got a clear spectroscopic signature of polariton formation from the cavity-coupled NO band of nitroprusside [13].

The study on the exciton-polariton dynamics modulated by exciton-photon detuning in a ZnO micro wire [14] is another important topics in this field. The investigation has shown that both the fast and slow decay rates of the exciton-polaritons are faster than that of the free exciton due to the strong coupling between excitons and photons in a micro cavity. The entropy of polariton states also has been studied [15]. The results show that due to the significant contribution of entropy on the free energy, spectroscopy does not properly arrange the free energy of the excited states. According to their free energy, the reordered states are useful to understand decoherence and to determine the potential of polariton states for reactivity.

To see the cavity molecular dynamics of polariton, the linear and non-linear response of liquid carbon dioxide under vibrational strong coupling conditions has been studied numerically [16]. The effect of multiple cavity modes on polariton relaxation is another interesting topic in this field [17]. But till now, it is not clear that how multiple cavity modes affect the polariton relaxation process. More realistic cavity description is needed to understand how cavities modify the molecular properties. The depolarization of optically confined semiconductor exciton-polariton condensates, polarization pinning and the optical orientation are other topics of investigation in recent years [18]. In case of the polariton dynamics in feedback-coupled cavities, it has been seen that a feedback cavity approach breaks the harmonic-oscillator restriction [19].

One interesting and applicable topic in this field is polariton laser. Several researchers are working in this field. To analyze the relaxation kinetics of cavity polaritons, the semiclassical Boltzmann equation can be used [20]. From this analysis, it is possible to know about the polariton laser and polariton superfluidity in microcavities. Polaritons go through Berezinskii Kosterlitz Thouless (BKT) phase transition towards superfluid phase and form a local quasi-condensation that allows polariton lasing for large band gap microcavities at temperatures over 300 K.

One of the difficulties of its development is that it has high fluctuation in intensity, therefore, low coherence, for which it is difficult to control its thermal stability especially for nanoscale devices. Another most important obstacle to get a polariton laser is the kinetic blocking or bottleneck effect of the relaxation of the polaritons. After the creation of excitons in the quantum well, at the time of thermalization, a depletion region is generated near the zone center that reduces the polaritons relaxation in that area. This effect is called bottleneck effect. It occurs because of thermalization as scattering with phonons is much slower as polaritons become more photon-like and recombination times are much shorter for photon-like polaritons. Because of an efficient electron-polariton scattering present in this case, this effect can be controlled by taking the low concentration of free electrons [20]. Polariton lasing at room temperature is one important topic within a polariton laser. The investigation is performed by using dielectric microcavity containing non-polar III-nitride quantum wells as an active medium [21]. It has been observed that microcavities exhibit two closely spaced lower polariton branches at room temperature. The electrically pumped inversion with less polariton lasing is also observed from a bulk GaN-based microcavity diode at room temperature [22]. A study on exciton-polariton lasing for topological defects at room temperature in an organic lattice was also performed [23]. Polariton lasing at an ultralow threshold, from a single GaN nanowire is another topic. It has been observed that the single GaN nanowire was strongly coupled with a large area of dielectric microcavity [24]. Polariton lasing can be obtained from a degenerate polariton condensate with non-resonant excitation by spontaneous radiative recombination [25]. Two-dimensional topological polariton laser is also an interesting topic in this field [26]. The investigation has also been performed on the semiconductor polariton laser which is a new source of coherent light [27]. This investigation will be helpful to get better polariton lasers and nonlinear polariton devices. The study on polariton laser by electrical pumping from microcavity which contains multiple quantum wells is one recent topic [28]. From the results it has been observed that the result will be helpful for practical implementation of polaritonic light source and by using wide band gap materials, it is possible to get it at room temperature. Polariton laser in the Bardeen-Cooper-Schrieffer Regime has also been studied [29].

But still, it is difficult to make a real polariton laser experimentally. Theoretically, to describe the formation of the coherent state and spin relaxation, a quantum theory should be developed. More clarification is needed on scattering in microcavities. Experimentally, it is important to fabricate a cavity that produces a strong coupling of exciton and light at room temperature. Increasing the polariton life time and improvement of mirror quality is also helpful.

BEC of polaritons in thermal equilibrium is another area in this field. Due to the leakage of microcavity, the polaritons' life time within microcavity is 30 ps or less [30]. At thermal equilibrium within an exciton polariton condensate, the role of quantum fluctuation has also been investigated [31]. A large phase window where both bosonic excitons and fermionic quasi-particles are present with strong coupling of photons is observed. The study [32] observes quantum depletion within a non-equilibrium exciton polariton condensate. It suggests more work needed to understand this field deeply.

There are few works that claim partial thermalization [33, 34], but it is difficult to get polariton condensate at thermal equilibrium due to the short lifetime of polariton. To get polaritons of long lifetime, a GaAs-based high-quality factor (Q) microcavity structure by molecular beam epitaxy may be helpful [35]. From the investigation, it has been observed that an improvement of the cavity Q by at least an order of magnitude is needed to get a polariton of long lifetime. This means the increase of one order of magnitude of the number of the quarter-wavelength layers in the distributed Bragg reflectors that make up the mirrors of the cavity is required. A polariton with a lifetime 270 ps at resonance is obtained by using a microcavity structure with the value of Q, which is approximately 320 000 and the value of cavity photon lifetime is approximately 135 ps [36].

Polariton emission in linear and nonlinear regime is also an important topic of research in this field [37-41]. The polariton emission dynamics directly depend on the polariton wave vector and exciton-photon energy detuning. When the excitonic and photonic wave functions are combined linearly, it produces the wave function of polariton which crucially depends on the bare excitonic and photonic energies. Since a photon-like polariton has a very small mass, approximately 10^{-4} of the mass of the electron, therefore, it has a very small density of states. Consequently, the probability of scattering of polariton by phonon is very low. The energy blue shift of polariton condensate arises due to the intermolecular energy change [41].

In the linear regime of polariton emission, three types of spectra were observed. First is the upper polariton branch, the second is the lower polariton branch and the third is the uncoupled (with photon) bound exciton [42]. The linear regime disappears with increasing pump power (excitation density threshold). To overcome this difficulty and to study the polariton dynamics in the linear regime, the excitation density has to be taken below its threshold value. Below the threshold, the relaxation directly depends upon the phonon scattering, which is not efficient to dissipate its linear regime.

The microscopic description of exciton-polaritons in microcavities is another important topic. It has been observed that the effect of particle exchange on the Rabi coupling is much smaller than expected in the standard treatment [43]. The effect of Pauli exclusion on Rabi coupling has been overestimated in the standard treatment. It means that the standard treatment neglected the light-induced modification of the exciton size.

Another topic in this field is the nonlinear effect in the spin relaxation of cavity polariton. Under the condition of pumping, it has been observed that the spin relaxation time as well as the polarization of dynamics of polariton is proportional to the detuning between the excitons and cavity mode energies. The positive detuning shows the curve in the middle of upper (UP) and lower polariton (LP) is parabolic. Typically, it means that the bottom of the photonic band lies above the excitonic band. Conversely, if the bottom of the excitonic band lies below the photonic band, it produces a non-parabolic curve.

Due to the variation of effective masses of the polaritons, a bottleneck region arises in the lower polariton branch. This reduces the polariton energy relaxation occurring by acoustic phonon scattering. There are two ways to overcome this difficulty. One is using doped samples with free carriers and another is using a strong pumping regime for which the polariton concentration will be high enough and polariton-polariton scattering will be efficient.

Collective dynamics of polariton condensate is another research area in this field. If the particle collapses with a common phase, it shows collective quantum behavior such as interference, quantized vortices and superfluidity. The behavior of super fluid motion of polaritons is different from BEC because BEC was only observed for dilute atomic gasses at small temperature in the range of μk.

The collective dynamics of polariton condensate in a semiconductor microcavity is another interesting area of research in this field and several works have been performed in this area [45-49]. From the results, diffusion less motion of polariton condensate is observed [46], a molecular superfluid phase exists with a regular polariton superfluid phase [47] and collective oscillations within a round box trap with low-energy and low-momentum of an exciton-polariton condensate is observed [49].

Many difficulties were observed by the researchers while studying the dynamics of polaritons. First, the polaritons' life-time is very short, a few picoseconds in the cavity (~ 4 ps). It hinders the detection of their dynamics and spatial nonhomogeneities provided by defects. Further, to detect their movement in the cavity, some difficulty arises due to the production of stray light from the resonant component. Finally, to block a stray light from a laser, a confocal setup is used in some experiments. Complexities were also observed due to the production of free electrons and holes as well as excitons simultaneously. It may have been controlled by varying the lattice temperature and population of photo-excited carriers.

Amplitude mode dynamics of polariton condensates is also an interesting topic. Superconductors and cold atomic gasses are very good examples of amplitude modes of polaritons. In amplitude modes, the density of condensed and uncondensed particles fluctuates. The Dicke model has been used to study amplitude mode dynamics of polariton condensate [50].

Upper polariton and lower polariton have been formed due to the strong coupling of excitons and photons in the semiconductor microcavity. It is possible to construct BEC states of polaritons as well as coherent light emission by the pumping in semiconductor microcavity. Outside the microcavity, the coupling of polariton with electromagnetic radiation may be coherently controlled by external pumping. With the excitation of phase mode, the collective behavior of microcavity polariton arises such as superfluidity and vortex dynamics. The amplitude mode of the upper polariton branch directly depends up the resonant excitation of the system.

The research on the amplitude mode in a microcavity polariton is another topic in this field. The amplitude mode is not stable due to two reasons: first, there is an interaction between amplitude modes and second, there is scattering between amplitude modes and phase mode. Due to these instabilities, an inhomogeneous condensate is formed [51]. When only the upper polariton is pumped, there is no instability due to wave mixing but due to attractive interactions. In this case, a condensate, actually a Bose supernova is produced for cold atomic gasses.

Polariton condensation within a semiconductor microcavity in the presence of a disorder is also an interesting topic. Sometimes, the disorder prevents to get polariton BEC [52]. It results in the dissipative nature of the exciton polaritons.

In a work [52], a particular defect potential (r) is applied that is composed of two Gaussian peaks of different amplitudes (see Fig. **5.5**). The longer one that is an extreme defect, located at the right hand side and the smaller one is located at the left-hand side that is indicated in the inset of (Fig. **3c**). When the pumping level is around the condensation threshold, due to this applied defect potential, a trapping potential in the middle is obtained that catches the condensate within it. In Fig. (**3**) **a**), a long pulse located within the pump spot appears after a significant delay. For higher pump pulse strength, polaritons emit in the negative direction that is in the small potential barrier region (see Fig. **3**, **b**, and **c**). Due to reflection and trapping potential, the multiple pulse polariton dynamics are observed. In the positive part of the space, a strong potential barrier triggers the irregular oscillations and blocks the flow. At higher pump pulse strength, the simulation results show that the multiple pulses merge into one fast polariton propagation of the polaritonic nonlinear wave in the negative direction.

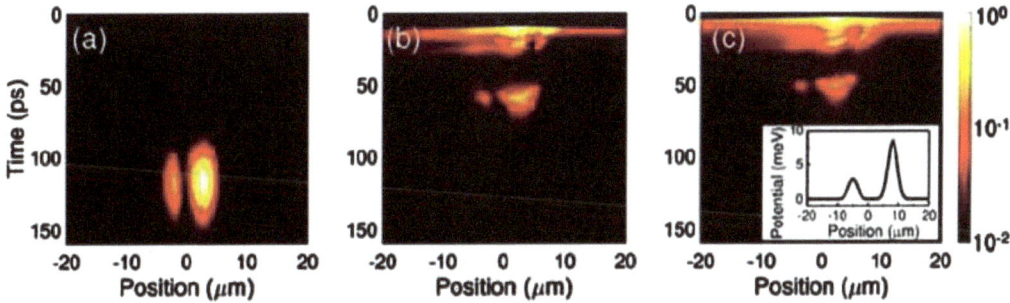

Fig. (5.5). Simulation results with disorder potential of a time resolved propagation of polariton condensate with different pump pulse strength. (**a**) Pump pulse strength 1.5 P_0 (**b**) Pump pulse strength 8 P_0 (**c**) Pump pulse strength 15 P_0. To observe the first condensation pulse, pumping amplitude denoted by P_0. The disorder potential is presented in the inset of (**c**). Here, the potential is represented with respect to position. Reproduced with permission from [52].

Sometimes, it is possible to get polariton BEC in the presence of a disorder [53]. Several works have been performed on the exciton-polariton condensate due to disorder [54-58]. It has been seen that the weak disorder acceleratescondensation as well as thermalization in several orders of magnitude [54].

Because of great scientific and technological significance, nowadays the study of the light propagation in a strongly correlated medium is very popular and increasing interest in both atomic and solid-state physics. Still, the effects of environmental coupling on the kinetics of single slow-light quanta are unknown. When the light propagates through an atomic BEC in the presence of strong interactions, polaron quasiparticles are formed. The effects of a strong interaction between the spin wave component and the surrounding condensate have been studied [59]. When the light couples with atoms weakly, then the polaron polariton emerges and it supports light propagation with spectral features. Dark state polaritons are formed due to photon dressing. It has been observed that polaritonic and polaronic quasiparticles are formed and investigation can be performed by the transmission spectrum of the interacting medium. It is a powerful setup for understanding the system with light-matter coupling in the presence of a strong interaction. This approach can be helpful to understand the other interaction effects in many body physics such as exciton polariton in semiconductors or interaction effect between atomic states. Several works have been done in this field [60 -62].

Polaritons in the organic system is a new and interesting topic in this field. Several works have been performed in this area [56, 63-65]. Hybrid organic inorganic polariton LED has been studied and a structure of strongly coupled two microcavities has been explained in one work [63]. One microcavity contains an organic semiconductor whereas another microcavity contains a series of weakly coupled inorganic QW. It has been reported that a delocalized polariton state is created by optical hybridization between the optical modes of the two cavities. This work gives a new idea to create a highly efficient polariton device. Due to strong light matter coupling, a new quantum state, organic cavity polariton is formed [64]. It has been claimed that using time-resolved microscopy, this organic exciton polariton induces long-range transport over several micrometers. In this case, the propagation velocity of polaritons is surprisingly lower than that expected. This work may be helpful in designing the future organic electronic devices. Organic exciton-polariton condensate in a lattice at room temperature also has been studied [65]. In this work, Frenkel excitons are created by fluorescent proteins and the soft nature of these Frenkel excitons allows making a photonic lattice. In a gap state, the self-localization of the condensate has also been investigated. The conclusion of this work is that a new organic polariton on GaAs platform may be produced, which has extensive application. The researchers of reference [23] also studied polariton lasers in an organic lattice at room temperature. They explained exciton-polariton lasing at room temperature for topological defects from the engraved lattice structure.

Fig. (**5.6 a, b, c**) represents the angle-resolved photoluminescence spectra from the domain boundary defect for increasing laser excitation power. Fig. (**5.6 d**) represents the integrated photoluminescence intensity and linewidth of the topological domain boundary defect with respect to excitation energy. At the polariton condensation threshold, for the value of $\approx 0.10 \, \text{nJ/pulse}$, line width decreases suddenly toward the resolution limit of the spectrometer which indicates the buildup of phase coherence [23]. Fig. (**5.6 e**) represents the variation in emission energy with laser excitation power. The output intensity increases sharply and nonlinearly by approximately 3 orders of magnitude. By measuring the correlation function, using a Michelson interferometer, the spatial and temporal coherence has been investigated [23].

Application of Bose Einstein Condensation of Polaritons

The investigation of the many body phenomena of Polaritons is not only due to scientific interest but also it has a lot of technological applications. Some optoelectronic devices like threshold lasers, ultrafast switches, and quantum simulators are made from the BEC of polariton.

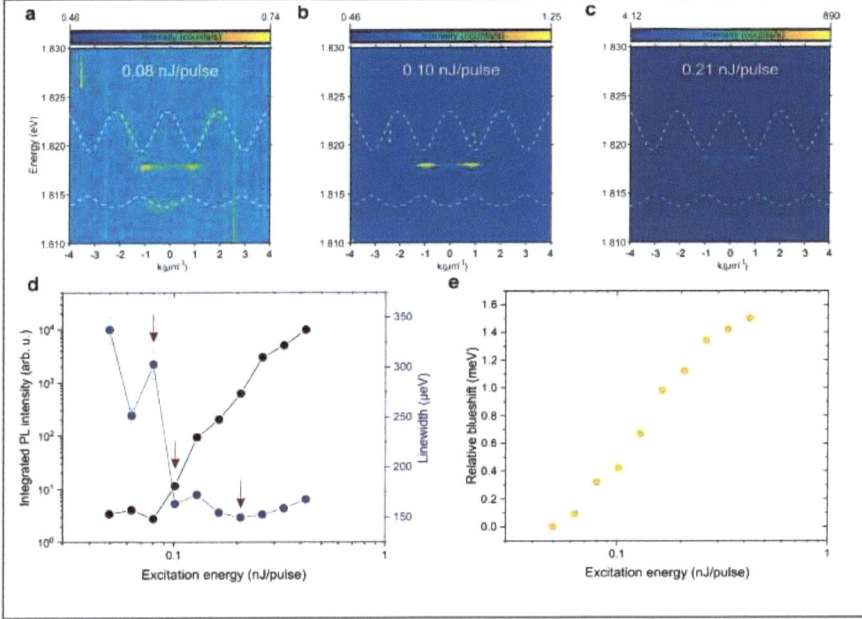

Fig. (5.6). (a, b, c) Angle-resolved photoluminescence spectra from the domain boundary defect for increasing laser excitation power (a) $P \approx 0.08$ nJ/pulse (b) $P \approx 0.10$ nJ/pulse (c) $P \approx 0.21$ nJ/pulse (d) The variation of linewidth and integrated photoluminescence intensity of the topological domain boundary defect with respect to excitation energy. The blue dotted line represents the linewidth and the black dotted line represents the integrated photoluminescence intensity. For the value of ≈ 0.10 nJ/pulse, at the polariton condensation threshold, a sudden decrease in the linewidth toward the resolution limit of the spectrometer accompanied by a strong nonlinear increase in the output intensity by approximately 3 orders of magnitude is observed. The three arrows indicate the dispersions plotted for 0.08 nJ/pulse, 0.10 nJ/pulse and 0.21 nJ/pulse, respectively. (e) Variation in emission energy with laser excitation power. A continuous increase of a blue shift (emission energy), induced by phase-space filling effects, with a total shift of 1.5 meV caused by the exciton-polariton nature of the system is observed. Reproduced with permission from [23].

Polariton Laser

Below the band inversion density *via* stimulated scattering into the condensate, Polariton lasers are formed. Therefore, polariton laser uses the benefits of ultralow threshold and response faster than the ordinary lasers that worked by band inversion. Polariton lasers go with spontaneous coherence and it has thresholds two or more orders of magnitude lower than that the ordinary photon lasers. A real polariton laser operates at room temperature, produced by electrical pumping. It maintains a strong coherence at room temperature even in the presence of thermal variation.

The study of polariton lasing with electrical insertion in GaAs microcavities at cryogenic temperatures was presented [25, 28]. But due to the small binding energy of GaAs, the operating temperature is limited. Recently polariton light emitting diodes (LED) and polariton laser have been studied in different systems [22, 66, 67, 63]. The coherent nature of the room temperature polariton lasers can be distinguished by calculating the first-order coherence function [68, 69, 70]. Research on the coherence of room temperature polariton lasers including the stability of intensity has been performed [27].

Polariton Switches

Microcavity exciton polariton is a mixed state of photon and exciton. It has properties like coherence and low dephasing of light as well as strong nonlinearity of excitons. It shows new light for next generation devices. Photonic information processing has many benefits over electronics, for example, low heating and high-speed operation. However, small nonlinearity in typical photonic systems and restriction in incorporation with electronic materials slow down its progress. Polariton switches respond very fast because of the ultrafast dynamics of polaritons. The fast movement of polaritons allows the switches to be nonlocal [71]. Based on the polariton bistability, the majority of polariton switches are made. The polariton population sharply increases when it is in resonance with the laser energy and gives rise to an S-shaped dependence of the polariton population as a function of power. At a low power, the transition takes place back to the low-population regime [72]. The work on the polariton bistable nonlinearity has been proposed in reference [73]. The study about polariton logic devices is explained in reference [74, 75]. The study by the researchers of reference [76] proposed a work by applying the property of having two stable states named spin bifurcation that does not depend on coherence and exists exterior of the system. An electrical spin switch was explained by the researcher [77]. Recently, a study by some researchers [78] explained optical spin switches built from the organic polaritons at room temperature.

Quantum States and Quantum Simulators

The quantum states of polaritons have also been investigated and suggested for future applications. The BEC of polaritons can be used in quantum simulators. The squeezed polariton states have been studied in reference [79]. The research on polariton antibunching has been reported [80, 81].

The field of quantum simulations established on the microcavity polaritons has been studied [82, 83]. Exciton polaritons have the ability to investigate the spin Hamiltonian and examine magnetic order by incorporating the spin dynamics manageable by light polarization. Recently some researchers reported [84] an electronically driven square and honeycomb lattice of exciton polaritons which showed the making of real polariton devices based on the polariton lattices for on-chip applications.

It is possible to simulate quantum fluids with different conditions and investigate quantum hydrodynamics for the polariton system. Within microcavity polariton systems, quantum turbulence has been studied by some researchers [85]. Within one-dimensional dark solitons, phase bistability has been studied [86]. Within a one-dimensional flat band, multistable dissipative gap, solitons have been studied [87]. A study proposed a theoretical strategy of quantum computing for the exciton polariton condensates created within the semiconductor micro pillars. [88]

In the pure photonic system, the effects of dissipation and dephasing are absent and therefore it can be used to make polariton quantum devices. Dissipation and dephasing can be stopped by symmetry-protected topological states that have been suggested by some researchers [89, 90] and observed in the polariton system [91].

CONCLUSION

Several experimental techniques have been used by several research groups to observe the BEC of polaritons. To investigate theoretically, the Gross-Pitaeveskii equation is useful to observe the BEC of polaritons. There are different excitation ways such as resonant pumping, non-resonant pumping, inhomogeneous optical pumping, *etc.* Real space imaging, momentum space imaging, spectroscopy, time resolved imaging are the few observation methods. This field is a growing field and it has lots of technological applications. Polariton laser, polariton switches, quantum states and quantum simulators are a few applications of it.

Though, research on polariton dynamics has already been conducted extensively more investigation is needed to get the answer to several unknown questions. The dynamics of polariton depend upon many parameters such as the radiative lifetime of polaritons, the excitation of the system, detuning between the excitons and cavity mode energies, as well as the relaxation towards its lowest state. Nowadays, one important thing is the polariton laser but still, it is difficult to get it in reality. More theoretical clarification is needed on scattering in microcavities. It is important to fabricate a cavity that produces a strong coupling of exciton and light at room temperature. Due to the short life span of polariton, it is difficult to get polariton condensate at thermal equilibrium. Nowadays, several research works are going on the organic polaritons. Research in this field may be helpful to create highly efficient polariton devices in future.

REFERENCES

[1]   Hayat, A.; Lange, C.; Rozema, L. A.; Chang, R.; Potnis, S.; van Driel, H. M.; Steinberg, A. M.; Steger, M.; Snoke, D. W.; Pfeifer, L. N.; West, K. W. Macroscopic coherence between quantum condensates formed at different times *Optics Express,* **2014,** *22*(25), 1-24.

[2]   Stevenson, R.; Astratov, V.; Skolnick, M.; Whittaker, D.; Emam-Ismail, M.; Tartakovskii, A.; Savvidis, P.; Baumberg, J.; Roberts, J. Continuous wave observation of massive polariton redistribution by stimulated scattering in semiconductor microcavities *Phys. Rev. Lett.,* **2000,** *85*(3), 3680.

        http://dx.doi.org/10.1103/PhysRevLett.85.3680

[3]   Langbein, W. Polariton correlation in microcavities produced by parametric scattering *Phys. Status Solidi B,* **2005,** *242*, 2260-2270.

        http://dx.doi.org/10.1002/pssb.200560967

[4]   Shannon, R.R. *Applied Optics and Optical Engineering;* Academic Press, **1980.**

[5]   Allen, L.; Beijersbergen, M.W.; Spreeuw, R.J.C.; Woerdman, J.P. Orbital angular momentum of light and the transformation of Laguerre-Gaussian laser modes. *Phys. Rev. A,* **1992,** *45*(11), 8185-8189.

        http://dx.doi.org/10.1103/PhysRevA.45.8185 PMID: 9906912

[6]   Nye, J.F.; Berry, M.V. Dislocations in wave trains. *Proc. R. Soc. Lond. A Math. Phys. Sci.,* **1974,** *336*(1605), 165-190.

        http://dx.doi.org/10.1098/rspa.1974.0012

[7]   Sanvitto, D.; Marchetti, F.M.; Szymańska, M.H.; Tosi, G.; Baudisch, M.; Laussy, F.P.; Krizhanovskii, D.N.; Skolnick, M.S.; Marrucci, L.; Lemaître, A.; Bloch, J.; Tejedor, C.; Viña, L. Persistent currents and quantized vortices in a polariton superfluid. *Nat. Phys.,* **2010,** *6*(7), 527-533.

        http://dx.doi.org/10.1038/nphys1668

[8]   Som, S.; Kieseling, F.; Stolz, H. Numerical simulation of exciton dynamics in Cu_2O at ultra-low temperatures within a potential trap. *J. Phys. Condens. Matter,* **2012,** *24*(33), 335803.

        http://dx.doi.org/10.1088/0953-8984/24/33/335803 PMID: 22836306

[9]   Deng, H.; Haug, H.; Yamamoto, Y. Exciton-polariton Bose-Einstein condensation. *Rev. Mod. Phys.,* **2010,** *82*(2), 1489-1537.

        http://dx.doi.org/10.1103/RevModPhys.82.1489

[10]   Richard, M.; Kasprzak, J.; Baas, A.; Kundermann, S.; Lagoudakis, K.G.; Wouters, M.; Carusotto, I.; Andre, R.; Pledran, B.D.; Dang, L.S. Exciton-polariton Bose-Einstein condensation: Advances and issues. *Int. J. Nanotechnol.,* **2010,** *7*(4/5/6/7/8), 668.
       http://dx.doi.org/10.1504/IJNT.2010.031738

[11]   Byrnes, T.; Kim, N.Y.; Yamamoto, Y. Exciton–polariton condensates. *Nat. Phys.,* **2014,** *10*(11), 803-813.
       http://dx.doi.org/10.1038/nphys3143

[12]   Hulkko, E.; Pikker, S.; Tiainen, V.; Tichauer, R.H.; Groenhof, G.; Toppari, J.J. Effect of molecular stokes shift on polariton dynamics. *J. Chem. Phys.,* **2021,** *154*(15), 154303.
       http://dx.doi.org/10.1063/5.0037896 PMID: 33887943

[13]   Grafton, A.B.; Dunkelberger, A.D.; Simpkins, B.S.; Triana, J.F.; Hernández, F.J.; Herrera, F.; Owrutsky, J.C. Excited-state vibration-polariton transitions and dynamics in nitroprusside. *Nat. Commun.,* **2021,** *12*(1), 214.
       http://dx.doi.org/10.1038/s41467-020-20535-z PMID: 33431901

[14]   Luo, S.; Wang, Y.; Liao, L.; Zhang, Z.; Shen, X.; Chen, Z. Exciton-polariton dynamics modulated by exciton-photon detuning in a ZnO microwire. *J. Appl. Phys.,* **2020,** *127*(2), 025702.
       http://dx.doi.org/10.1063/1.5133005

[15]   Scholes, G.D.; DelPo, C.A.; Kudisch, B. Entropy reorders polariton states. *J. Phys. Chem. Lett.,* **2020,** *11*(15), 6389-6395.
       http://dx.doi.org/10.1021/acs.jpclett.0c02000 PMID: 32678609

[16]   Li, T.E.; Nitzan, A.; Subotnik, J.E. Cavity molecular dynamics simulations of vibrational polariton-enhanced molecular nonlinear absorption. *J. Chem. Phys.,* **2021,** *154*(9), 094124.
       http://dx.doi.org/10.1063/5.0037623 PMID: 33685184

[17]   Tichauer, R.H.; Feist, J.; Groenhof, G. Multi-scale dynamics simulations of molecular polaritons: The effect of multiple cavity modes on polariton relaxation. *J. Chem. Phys.,* **2021,** *154*(10), 104112.
       http://dx.doi.org/10.1063/5.0037868 PMID: 33722041

[18]   Gnusov, I.; Sigurdsson, H.; Baryshev, S.; Ermatov, T.; Askitopoulos, A.; Lagoudakis, P.G. Optical orientation, polarization pinning, and depolarization dynamics in optically confined polariton condensates. *Phys. Rev. B,* **2020,** *102*(12), 125419.
       http://dx.doi.org/10.1103/PhysRevB.102.125419

[19]   Yao, B.; Gui, Y.S.; Rao, J.W.; Kaur, S.; Chen, X.S.; Lu, W.; Xiao, Y.; Guo, H.; Marzlin, K.P.; Hu, C.M. Cooperative polariton dynamics in feedback-coupled cavities. *Nat. Commun.,* **2017,** *8*(1), 1437.
       http://dx.doi.org/10.1038/s41467-017-01796-7 PMID: 29127391

[20]   Kavokin, A.; Malpuech, G.; Laussy, F.P. Polariton laser and polariton superfluidity in microcavities. *Phys. Lett. A,* **2003,** *306*(4), 187-199.
       http://dx.doi.org/10.1016/S0375-9601(02)01579-7

[21]   Amargianitakis, E.A.; Tsagaraki, K.; Kostopoulos, A.; Konstantinidis, G.; Delamadeleine, E.; Monroy, E.; Pelekanos, N.T. Non-polar GaN/AlGaN quantum-well polariton laser at room temperature. *Phys. Rev. B,* **2021,** *104*(12), 125311.

http://dx.doi.org/10.1103/PhysRevB.104.125311

[22] Bhattacharya, P.; Frost, T.; Deshpande, S.; Baten, M.Z.; Hazari, A.; Das, A. Room temperature electrically injected polariton laser. *Phys. Rev. Lett.,* **2014,** *112*(23), 236802.
http://dx.doi.org/10.1103/PhysRevLett.112.236802 PMID: 24972222

[23] Dusel, M.; Betzold, S.; Harder, T.H.; Emmerling, M.; Beierlein, J.; Ohmer, J.; Fischer, U.; Thomale, R.; Schneider, C.; Höfling, S.; Klembt, S. Room-temperature topological polariton laser in an organic lattice. *Nano Lett.,* **2021,** *21*(15), 6398-6405.
http://dx.doi.org/10.1021/acs.nanolett.1c00661 PMID: 34328737

[24] Das, A.; Heo, J.; Jankowski, M.; Guo, W.; Zhang, L.; Deng, H.; Bhattacharya, P. Room temperature ultralow threshold GaN nanowire polariton laser. *Phys. Rev. Lett.,* **2011,** *107*(6), 066405.
http://dx.doi.org/10.1103/PhysRevLett.107.066405 PMID: 21902349

[25] Bhattacharya, P.; Xiao, B.; Das, A.; Bhowmick, S.; Heo, J. Solid state electrically injected exciton-polariton laser. *Phys. Rev. Lett.,* **2013,** *110*(20), 206403.
http://dx.doi.org/10.1103/PhysRevLett.110.206403 PMID: 25167434

[26] Kartashov, Y.V.; Skryabin, D.V. Two-dimensional topological polariton laser. *Phys. Rev. Lett.,* **2019,** *122*(8), 083902.
http://dx.doi.org/10.1103/PhysRevLett.122.083902 PMID: 30932611

[27] Kim, S.; Zhang, B.; Wang, Z.; Fischer, J.; Brodbeck, S.; Kamp, M.; Schneider, C.; Höfling, S.; Deng, H. Coherent polariton laser. *Phys. Rev. X,* **2016,** *6*(1), 011026.
http://dx.doi.org/10.1103/PhysRevX.6.011026

[28] Schneider, C.; Rahimi-Iman, A.; Kim, N.Y.; Fischer, J.; Savenko, I.G.; Amthor, M.; Lermer, M.; Wolf, A.; Worschech, L.; Kulakovskii, V.D.; Shelykh, I.A.; Kamp, M.; Reitzenstein, S.; Forchel, A.; Yamamoto, Y.; Höfling, S. An electrically pumped polariton laser. *Nature,* **2013,** *497*(7449), 348-352.
http://dx.doi.org/10.1038/nature12036 PMID: 23676752

[29] Hu, J.; Wang, Z.; Kim, S.; Deng, H.; Brodbeck, S.; Schneider, C.; Höfling, S.; Kwong, N.H.; Binder, R. Polariton laser in the bardeen-cooper-schrieffer regime. *Phys. Rev. X,* **2021,** *11*(1), 011018.
http://dx.doi.org/10.1103/PhysRevX.11.011018

[30] Wertz, E.; Ferrier, L.; Solnyshkov, D.D.; Johne, R.; Sanvitto, D.; Lemaître, A.; Sagnes, I.; Grousson, R.; Kavokin, A.V.; Senellart, P.; Malpuech, G.; Bloch, J. Spontaneous formation and optical manipulation of extended polariton condensates. *Nat. Phys.,* **2010,** *6*(11), 860-864.
http://dx.doi.org/10.1038/nphys1750

[31] Hu, H.; Liu, X.J. Quantum fluctuations in a strongly interacting bardeen-cooper-schrieffer polariton condensate at thermal equilibrium. *Phys. Rev. A (Coll. Park),* **2020,** *101*(1), 011602.
http://dx.doi.org/10.1103/PhysRevA.101.011602

[32] Pieczarka, M.; Estrecho, E.; Boozarjmehr, M.; Bleu, O.; Steger, M.; West, K.; Pfeiffer, L.N.; Snoke, D.W.; Levinsen, J.; Parish, M.M.; Truscott, A.G.; Ostrovskaya, E.A. Observation of quantum depletion in a non-equilibrium exciton–polariton condensate. *Nat. Commun.,* **2020,** *11*(1), 429.

http://dx.doi.org/10.1038/s41467-019-14243-6 PMID: 31969565

[33]   Deng, H.; Press, D.; Götzinger, S.; Solomon, G.S.; Hey, R.; Ploog, K.H.; Yamamoto, Y. Quantum degenerate exciton-polaritons in thermal equilibrium. *Phys. Rev. Lett.,* **2006,** *97*(14), 146402.
http://dx.doi.org/10.1103/PhysRevLett.97.146402 PMID: 17155273

[34]   Kasprzak, J.; Solnyshkov, D.D.; André, R.; Dang, L.S.; Malpuech, G. Formation of an exciton polariton condensate: Thermodynamic *versus* kinetic regimes. *Phys. Rev. Lett.,* **2008,** *101*(14), 146404.
http://dx.doi.org/10.1103/PhysRevLett.101.146404 PMID: 18851551

[35]   Sun, Y.; Wen, P.; Yoon, Y.; Liu, G.; Steger, M.; Pfeiffer, L.N.; West, K.; Snoke, D.W.; Nelson, K.A. Bose-einstein condensation of long-lifetime polaritons in thermal equilibrium. *Phys. Rev. Lett.,* **2017,** *118*(1), 016602.
http://dx.doi.org/10.1103/PhysRevLett.118.016602 PMID: 28106443

[36]   Sun, Y.; Wen, P.; Yoon, Y.; Liu, G.; Steger, M.; Pfeiffer, L.N.; West, K.; Snoke, D.W.; Nelson, K.A. Erratum: Bose-einstein condensation of long-lifetime polaritons in thermal equilibrium. *Phys. Rev. Lett.,* **2017,** *118*(14), 149901.
http://dx.doi.org/10.1103/PhysRevLett.118.149901 PMID: 28430509

[37]   Anton-Solanas, C.; Waldherr, M.; Klaas, M.; Suchomel, H.; Harder, T.H.; Cai, H.; Sedov, E.; Klembt, S.; Kavokin, A.V.; Tongay, S.; Watanabe, K.; Taniguchi, T.; Höfling, S.; Schneider, C. Bosonic condensation of exciton–polaritons in an atomically thin crystal. *Nat. Mater.,* **2021,** *20*(9), 1233-1239.
http://dx.doi.org/10.1038/s41563-021-01000-8 PMID: 33958772

[38]   Polimeno, L.; Fieramosca, A.; Lerario, G.; Cinquino, M.; De Giorgi, M.; Ballarini, D.; Todisco, F.; Dominici, L.; Ardizzone, V.; Pugliese, M.; Prontera, C.T.; Maiorano, V.; Gigli, G.; De Marco, L.; Sanvitto, D. Observation of two thresholds leading to polariton condensation in 2D hybrid perovskites. *Adv. Opt. Mater.,* **2020,** *8*(16), 2000176.
http://dx.doi.org/10.1002/adom.202000176

[39]   Wang, J.; Xu, H.; Su, R.; Peng, Y.; Wu, J.; Liew, T.C.H.; Xiong, Q. Spontaneously coherent orbital coupling of counterrotating exciton polaritons in annular perovskite microcavities. *Light Sci. Appl.,* **2021,** *10*(1), 45.
http://dx.doi.org/10.1038/s41377-021-00478-w PMID: 33649295

[40]   Wu, J.; Ghosh, S.; Su, R.; Fieramosca, A.; Liew, T.C.H.; Xiong, Q. Nonlinear parametric scattering of exciton polaritons in perovskite microcavities. *Nano Lett.,* **2021,** *21*(7), 3120-3126.
http://dx.doi.org/10.1021/acs.nanolett.1c00283 PMID: 33788571

[41]   Yagafarov, T.; Sannikov, D.; Zasedatelev, A.; Georgiou, K.; Baranikov, A.; Kyriienko, O.; Shelykh, I.; Gai, L.; Shen, Z.; Lidzey, D.; Lagoudakis, P. Mechanisms of blueshifts in organic polariton condensates. *Commun. Phys.,* **2020,** *3*(1), 18.
http://dx.doi.org/10.1038/s42005-019-0278-6

[42]   Kłopotowski, Ł.; Santos, R.; Amo, A.; Martín, M.D.; Viña, L.; André, R. Dynamics of polariton emission in the linear regime. *Acta Phys. Pol. A,* **2004,** *106*(3), 443-450.
http://dx.doi.org/10.12693/APhysPolA.106.443

[43] Levinsen, J.; Li, G.; Parish, M.M. Microscopic description of exciton-polaritons in microcavities. *Phys. Rev. Res.*, **2019**, *1*(3), 033120.
      http://dx.doi.org/10.1103/PhysRevResearch.1.033120

[44] Solnyshkov, D.D.; Shelykh, I.A.; Glazov, M.M.; Malpuech, G.; Amand, T.; Renucci, P.; Marie, X.; Kavokin, A.V. Nonlinear effects in spin relaxation of cavity polaritons. *Semiconductors*, **2007**, *41*(9), 1080-1091.
      http://dx.doi.org/10.1134/S1063782607090138

[45] Amo, A.; Sanvitto, D.; Viña, L. Collective dynamics of excitons and polaritons in semiconductor nanostructures. *Semicond. Sci. Technol.*, **2010**, *25*(4), 043001.
      http://dx.doi.org/10.1088/0268-1242/25/4/043001

[46] Amo, A.; Sanvitto, D.; Laussy, F.P.; Ballarini, D.; Valle, E.; Martin, M.D.; Lemaître, A.; Bloch, J.; Krizhanovskii, D.N.; Skolnick, M.S.; Tejedor, C.; Viña, L. Collective fluid dynamics of a polariton condensate in a semiconductor microcavity. *Nature*, **2009**, *457*(7227), 291-295.
      http://dx.doi.org/10.1038/nature07640 PMID: 19148095

[47] Marchetti, F.M.; Keeling, J. Collective pairing of resonantly coupled microcavity polaritons. *Phys. Rev. Lett.*, **2014**, *113*(21), 216405.
      http://dx.doi.org/10.1103/PhysRevLett.113.216405 PMID: 25479511

[48] Biegańska, D.; Pieczarka, M.; Estrecho, E.; Steger, M.; Snoke, D.W.; West, K.; Pfeiffer, L.N.; Syperek, M.; Truscott, A.G.; Ostrovskaya, E.A. Collective excitations of exciton-polariton condensates in a synthetic gauge field. *Phys. Rev. Lett.*, **2021**, *127*(18), 185301.
      http://dx.doi.org/10.1103/PhysRevLett.127.185301 PMID: 34767383

[49] Estrecho, E.; Pieczarka, M.; Wurdack, M.; Steger, M.; West, K.; Pfeiffer, L.N.; Snoke, D.W.; Truscott, A.G.; Ostrovskaya, E.A. Low-energy collective oscillations and bogoliubov sound in an exciton-polariton condensate. *Phys. Rev. Lett.*, **2021**, *126*(7), 075301.
      http://dx.doi.org/10.1103/PhysRevLett.126.075301 PMID: 33666453

[50] Brierley, R.T.; Littlewood, P.B.; Eastham, P.R. Amplitude-mode dynamics of polariton condensates. *Phys. Rev. Lett.*, **2011**, *107*(4), 040401.
      http://dx.doi.org/10.1103/PhysRevLett.107.040401 PMID: 21866986

[51] Steger, M.; Hanai, R.; Edelman, A.O.; Littlewood, P.B.; Snoke, D.W.; Beaumariage, J.; Fluegel, B.; West, K.; Pfeiffer, L.N.; Mascarenhas, A. Direct observation of the quantum fluctuation driven amplitude mode in a microcavity polariton condensate. *Phys. Rev. B*, **2021**, *103*(20), 205125.
      http://dx.doi.org/10.1103/PhysRevB.103.205125

[52] Pieczarka, M.; Syperek, M.; Dusanowski, Ł.; Opala, A.; Langer, F.; Schneider, C.; Höfling, S.; Sęk, G. Relaxation oscillations and ultrafast emission pulses in a disordered expanding polariton condensate. *Sci. Rep.*, **2017**, *7*(1), 7094.
      http://dx.doi.org/10.1038/s41598-017-07470-8 PMID: 28769102

[53] Baas, A.; Lagoudakis, K.G.; Richard, M.; André, R.; Dang, L.S.; Deveaud-Plédran, B. Synchronized and desynchronized phases of exciton-polariton condensates in the presence of disorder. *Phys. Rev. Lett.*, **2008**, *100*(17), 170401.
      http://dx.doi.org/10.1103/PhysRevLett.100.170401 PMID: 18518258

[54]   Fusaro, A.; Garnier, J.; Krupa, K.; Millot, G.; Picozzi, A. Dramatic acceleration of wave condensation mediated by disorder in multimode fibers. *Phys. Rev. Lett.,* **2019,** *122*(12), 123902.

http://dx.doi.org/10.1103/PhysRevLett.122.123902 PMID: 30978031

[55]   Polak, D.; Jayaprakash, R.; Lyons, T.P.; Martínez-Martínez, L.Á.; Leventis, A.; Fallon, K.J.; Coulthard, H.; Bossanyi, D.G.; Georgiou, K.; Petty, A.J., II; Anthony, J.; Bronstein, H.; Yuen-Zhou, J.; Tartakovskii, A.I.; Clark, J.; Musser, A.J. Manipulating molecules with strong coupling: Harvesting triplet excitons in organic exciton microcavities. *Chem. Sci. (Camb.),* **2020,** *11*(2), 343-354.

http://dx.doi.org/10.1039/C9SC04950A PMID: 32190258

[56]   Keeling, J.; Kéna-Cohen, S. Bose–einstein condensation of exciton-polaritons in organic microcavities. *Annu. Rev. Phys. Chem.,* **2020,** *71*(1), 435-459.

http://dx.doi.org/10.1146/annurev-physchem-010920-102509 PMID: 32126177

[57]   Ren, J.; Liao, Q.; Huang, H.; Li, Y.; Gao, T.; Ma, X.; Schumacher, S.; Yao, J.; Bai, S.; Fu, H. Efficient bosonic condensation of exciton polaritons in an H-aggregate organic single-crystal microcavity. *Nano Lett.,* **2020,** *20*(10), 7550-7557.

http://dx.doi.org/10.1021/acs.nanolett.0c03009 PMID: 32986448

[58]   Ballarini, D.; Chestnov, I.; Caputo, D.; De Giorgi, M.; Dominici, L.; West, K.; Pfeiffer, L.N.; Gigli, G.; Kavokin, A.; Sanvitto, D. Self-trapping of exciton-polariton condensates in GaAs microcavities. *Phys. Rev. Lett.,* **2019,** *123*(4), 047401.

http://dx.doi.org/10.1103/PhysRevLett.123.047401 PMID: 31491238

[59]   Camacho-Guardian, A.; Nielsen, K.K.; Pohl, T.; Bruun, G.M. Polariton dynamics in strongly interacting quantum many-body systems. *Phys. Rev. Res.,* **2020,** *2*(2), 023102.

http://dx.doi.org/10.1103/PhysRevResearch.2.023102

[60]   Das, A.; Bhattacharya, P.; Heo, J.; Banerjee, A.; Guo, W. Polariton Bose–Einstein condensate at room temperature in an Al(Ga)N nanowire–dielectric microcavity with a spatial potential trap. *Proc. Natl. Acad. Sci. USA,* **2013,** *110*(8), 2735-2740.

http://dx.doi.org/10.1073/pnas.1210842110 PMID: 23382183

[61]   Roux, K.; Konishi, H.; Helson, V.; Brantut, J.P. Strongly correlated Fermions strongly coupled to light. *Nat. Commun.,* **2020,** *11*(1), 2974.

http://dx.doi.org/10.1038/s41467-020-16767-8 PMID: 32532985

[62]   Cantu, S.H.; Venkatramani, A.V.; Xu, W.; Zhou, L.; Jelenković, B.; Lukin, M.D.; Vuletić, V. Repulsive photons in a quantum nonlinear medium. *Nat. Phys.,* **2020,** *16*(9), 921-925.

http://dx.doi.org/10.1038/s41567-020-0917-6

[63]   Jayaprakash, R.; Georgiou, K.; Coulthard, H.; Askitopoulos, A.; Rajendran, S.K.; Coles, D.M.; Musser, A.J.; Clark, J.; Samuel, I.D.W.; Turnbull, G.A.; Lagoudakis, P.G.; Lidzey, D.G. A hybrid organic–inorganic polariton LED. *Light Sci. Appl.,* **2019,** *8*(1), 81.

http://dx.doi.org/10.1038/s41377-019-0180-8 PMID: 31666947

[64]   Rozenman, G.G.; Akulov, K.; Golombek, A.; Schwartz, T. Long-range transport of organic exciton-polaritons revealed by ultrafast microscopy. *ACS Photonics,* **2018,** *5*(1), 105-110.

http://dx.doi.org/10.1021/acsphotonics.7b01332

[65]   Dusel, M.; Betzold, S.; Egorov, O.A.; Klembt, S.; Ohmer, J.; Fischer, U.; Höfling, S.; Schneider, C. Room temperature organic exciton–polariton condensate in a lattice. *Nat. Commun.,* **2020,** *11*(1), 2863.
       http://dx.doi.org/10.1038/s41467-020-16656-0 PMID: 32514026

[66]   Graf, A.; Held, M.; Zakharko, Y.; Tropf, L.; Gather, M.C.; Zaumseil, J. Electrical pumping and tuning of exciton-polaritons in carbon nanotube microcavities. *Nat. Mater.,* **2017,** *16*(9), 911-917.
       http://dx.doi.org/10.1038/nmat4940 PMID: 28714985

[67]   Gu, J.; Chakraborty, B.; Khatoniar, M.; Menon, V.M. A room-temperature polariton light-emitting diode based on monolayer WS$_2$. *Nat. Nanotechnol.,* **2019,** *14*(11), 1024-1028.
       http://dx.doi.org/10.1038/s41565-019-0543-6 PMID: 31548689

[68]   Christopoulos, S.; von Högersthal, G.B.H.; Grundy, A.J.D.; Lagoudakis, P.G.; Kavokin, A.V.; Baumberg, J.J.; Christmann, G.; Butté, R.; Feltin, E.; Carlin, J.F.; Grandjean, N. Room-temperature polariton lasing in semiconductor microcavities. *Phys. Rev. Lett.,* **2007,** *98*(12), 126405.
       http://dx.doi.org/10.1103/PhysRevLett.98.126405 PMID: 17501142

[69]   Daskalakis, K.S.; Maier, S.A.; Murray, R.; Kéna-Cohen, S. Nonlinear interactions in an organic polariton condensate. *Nat. Mater.,* **2014,** *13*(3), 271-278.
       http://dx.doi.org/10.1038/nmat3874 PMID: 24509602

[70]   Wang, Z.; Zhang, B.; Deng, H. Dispersion engineering for vertical microcavities using subwavelength gratings. *Phys. Rev. Lett.,* **2015,** *114*(7), 073601.
       http://dx.doi.org/10.1103/PhysRevLett.114.073601 PMID: 25763957

[71]   Amo, A.; Liew, T.C.H.; Adrados, C.; Houdré, R.; Giacobino, E.; Kavokin, A.V.; Bramati, A. Exciton–polariton spin switches. *Nat. Photonics,* **2010,** *4*(6), 361-366.
       http://dx.doi.org/10.1038/nphoton.2010.79

[72]   Baas, A.; Karr, J.P.; Eleuch, H.; Giacobino, E. Optical bistability in semiconductor microcavities. *Phys. Rev. A,* **2004,** *69*(2), 023809.
       http://dx.doi.org/10.1103/PhysRevA.69.023809

[73]   Liew, T.C.H.; Kavokin, A.V.; Shelykh, I.A. Optical circuits based on polariton neurons in semiconductor microcavities. *Phys. Rev. Lett.,* **2008,** *101*(1), 016402.
       http://dx.doi.org/10.1103/PhysRevLett.101.016402 PMID: 18764129

[74]   Amo, A.; Pigeon, S.; Adrados, C.; Houdré, R.; Giacobino, E.; Ciuti, C.; Bramati, A. Light engineering of the polariton landscape in semiconductor microcavities. *Phys. Rev. B Condens. Matter Mater. Phys.,* **2010,** *82*(8), 081301.
       http://dx.doi.org/10.1103/PhysRevB.82.081301

[75]   Ballarini, D.; De Giorgi, M.; Cancellieri, E.; Houdré, R.; Giacobino, E.; Cingolani, R.; Bramati, A.; Gigli, G.; Sanvitto, D. All-optical polariton transistor. *Nat. Commun.,* **2013,** *4*(1), 1778.
       http://dx.doi.org/10.1038/ncomms2734 PMID: 23653190

[76]   Ohadi, H.; Dreismann, A.; Rubo, Y.G.; Pinsker, F.; del Valle-Inclan Redondo, Y.; Tsintzos, S.I.; Hatzopoulos, Z.; Savvidis, P.G.; Baumberg, J.J. Spontaneous spin bifurcations and

ferromagnetic phase transitions in a spinor exciton-polariton condensate. *Phys. Rev. X,* **2015,** *5*(3), 031002.

http://dx.doi.org/10.1103/PhysRevX.5.031002

[77]   Dreismann, A.; Ohadi, H.; del Valle-Inclan Redondo, Y.; Balili, R.; Rubo, Y.G.; Tsintzos, S.I.; Deligeorgis, G.; Hatzopoulos, Z.; Savvidis, P.G.; Baumberg, J.J. A sub-femtojoule electrical spin-switch based on optically trapped polariton condensates. *Nat. Mater.,* **2016,** *15*(10), 1074-1078.

http://dx.doi.org/10.1038/nmat4722 PMID: 27500807

[78]   Zasedatelev, A.V.; Baranikov, A.V.; Urbonas, D.; Scafirimuto, F.; Scherf, U.; Stöferle, T.; Mahrt, R.F.; Lagoudakis, P.G. A room-temperature organic polariton transistor. *Nat. Photonics,* **2019,** *13*(6), 378-383.

http://dx.doi.org/10.1038/s41566-019-0392-8

[79]   Boulier, T.; Bamba, M.; Amo, A.; Adrados, C.; Lemaitre, A.; Galopin, E.; Sagnes, I.; Bloch, J.; Ciuti, C.; Giacobino, E.; Bramati, A. Polariton-generated intensity squeezing in semiconductor micropillars. *Nat. Commun.,* **2014,** *5*(1), 3260.

http://dx.doi.org/10.1038/ncomms4260 PMID: 24518009

[80]   Delteil, A.; Fink, T.; Schade, A.; Höfling, S.; Schneider, C.; İmamoğlu, A. Towards polariton blockade of confined exciton–polaritons. *Nat. Mater.,* **2019,** *18*(3), 219-222.

http://dx.doi.org/10.1038/s41563-019-0282-y PMID: 30783230

[81]   Muñoz-Matutano, G.; Wood, A.; Johnsson, M.; Vidal, X.; Baragiola, B.Q.; Reinhard, A.; Lemaître, A.; Bloch, J.; Amo, A.; Nogues, G.; Besga, B.; Richard, M.; Volz, T. Emergence of quantum correlations from interacting fibre-cavity polaritons. *Nat. Mater.,* **2019,** *18*(3), 213-218.

http://dx.doi.org/10.1038/s41563-019-0281-z PMID: 30783231

[82]   Berloff, N.G.; Silva, M.; Kalinin, K.; Askitopoulos, A.; Töpfer, J.D.; Cilibrizzi, P.; Langbein, W.; Lagoudakis, P.G. Realizing the classical XY Hamiltonian in polariton simulators. *Nat. Mater.,* **2017,** *16*(11), 1120-1126.

http://dx.doi.org/10.1038/nmat4971 PMID: 28967915

[83]   Lai, C.W.; Kim, N.Y.; Utsunomiya, S.; Roumpos, G.; Deng, H.; Fraser, M.D.; Byrnes, T.; Recher, P.; Kumada, N.; Fujisawa, T.; Yamamoto, Y. Coherent zero-state and π-state in an exciton–polariton condensate array. *Nature,* **2007,** *450*(7169), 529-532.

http://dx.doi.org/10.1038/nature06334 PMID: 18033292

[84]   Suchomel, H.; Klembt, S.; Harder, T.H.; Klaas, M.; Egorov, O.A.; Winkler, K.; Emmerling, M.; Thomale, R.; Höfling, S.; Schneider, C. Platform for electrically pumped polariton simulators and topological lasers. *Phys. Rev. Lett.,* **2018,** *121*(25), 257402.

http://dx.doi.org/10.1103/PhysRevLett.121.257402 PMID: 30608796

[85]   Boulier, T.; Cancellieri, E.; Sangouard, N.D.; Glorieux, Q.; Kavokin, A.V.; Whittaker, D.M.; Giacobino, E.; Bramati, A. Injection of orbital angular momentum and storage of quantized vortices in polariton superfluids. *Phys. Rev. Lett.,* **2016,** *116*(11), 116402.

http://dx.doi.org/10.1103/PhysRevLett.116.116402 PMID: 27035313

[86]   Goblot, V.; Nguyen, H.S.; Carusotto, I.; Galopin, E.; Lemaître, A.; Sagnes, I.; Amo, A.; Bloch, J. Phase-controlled bistability of a dark soliton train in a polariton fluid. *Phys. Rev. Lett.,* **2016**, *117*(21), 217401.

http://dx.doi.org/10.1103/PhysRevLett.117.217401 PMID: 27911548

[87]   Goblot, V.; Rauer, B.; Vicentini, F.; Le Boité, A.; Galopin, E.; Lemaître, A.; Le Gratiet, L.; Harouri, A.; Sagnes, I.; Ravets, S.; Ciuti, C.; Amo, A.; Bloch, J. Nonlinear polariton fluids in a flatband reveal discrete gap solitons. *Phys. Rev. Lett.,* **2019**, *123*(11), 113901.

http://dx.doi.org/10.1103/PhysRevLett.123.113901 PMID: 31573264

[88]   Ghosh, S.; Liew, T. C. H. Quantum computing with exciton-polariton condensates. *npj Quantum Inf.,* **2020**, *6*, 16.

http://dx.doi.org/10.1038/s41534-020-0244-x

[89]   Karzig, T.; Bardyn, C.E.; Lindner, N.H.; Refael, G. Topological polaritons. *Phys. Rev. X,* **2015**, *5*(3), 031001.

http://dx.doi.org/10.1103/PhysRevX.5.031001

[90]   Nalitov, A.V.; Solnyshkov, D.D.; Malpuech, G. Polariton Z topological insulator. *Phys. Rev. Lett.,* **2015**, *114*(11), 116401.

http://dx.doi.org/10.1103/PhysRevLett.114.116401 PMID: 25839295

[91]   Klembt, S.; Harder, T.H.; Egorov, O.A.; Winkler, K.; Ge, R.; Bandres, M.A.; Emmerling, M.; Worschech, L.; Liew, T.C.H.; Segev, M.; Schneider, C.; Höfling, S. Exciton-polariton topological insulator. *Nature,* **2018**, *562*(7728), 552-556.

http://dx.doi.org/10.1038/s41586-018-0601-5 PMID: 30297800

SUBJECT INDEX

A

Absorption imaging 78

Angular momentum 91, 96, 98
  orbital 96
  spin 98

Angular velocity 58

Annihilation 44, 75, 76
  operator 44
  suppressed 76

Applications 16, 17, 58, 63, 73, 75, 76, 77, 81, 82, 83, 84, 85, 91, 106, 113, 115, 116
  of bose einstein condensation of excitons 77
  of gravity sensor 84
  of magnetic sensors 83
  of polariton condensation 91
  of Rotation Sensor 85
  on-chip 116
  optoelectronic 75, 76
  industrial 76
  technological 63, 91, 106, 113, 116

Applying uniaxial stress 64

Atom(s) 2, 4, 16, 22, 39, 48, 73, 77, 78, 79, 80, 84, 112
  density distributions of atom lasers 78
  hydrogen 22, 39
  laser 16, 77, 78
  transition metal 73

Atomic 12, 58, 78, 79, 110
  gasses 58, 110
  masses 12
  oscillations 79
  wave packet 78

Auger 26, 32, 34
  decay process 32
  process 34
  recombination 26

B

Back-reflection geometry 96

Band(s) 19, 20, 24, 26, 42, 43, 114
  electronic energy 24
  gap energy 19, 20, 26, 42, 43
  inversion density 114

BEC 1, 2, 16, 17, 58, 63, 64, 76, 81, 84, 85, 91, 96, 111, 112, 114, 115, 116
  gas 64
  gravimeter 84
  of bosons 1
  of excitons 16, 63, 85
  of excitons and polaritons 17
  of polaritons 17, 58, 91, 96, 111, 112, 114, 115, 116
  of rubidium atoms 76
  sensor 81
  theoretical description of 1, 2

Binding energy 15, 21, 22, 23, 40, 41, 42

Boltzmann 15, 23, 24, 25, 33, 42, 69, 72
  distribution 42
  equation 15, 23, 24, 25, 33, 69, 72

Bose 48, 63, 64
  condensates 48
  -Einstein distribution 63, 64

Bose Einstein condensation (BEC) 1, 7, 11, 15, 16, 17, 43, 44, 48, 63, 64, 65, 76, 84, 91, 110
  dynamical 48
  of atoms 48

Boson(s) 1, 2, 4, 13, 25, 73, 91, 104, 105
  composite 13
  gas 25
  interaction 104

Bosonic 36, 49
  quantum degeneracy 49
  quasiparticle 36

Bragg polaritons 36

C

Cavity 39, 76
  microribbon 76
  photons 39

www.ingramcontent.com/pod-product-compliance
Lightning Source LLC
Chambersburg PA
CBHW041715210326

41598CB00007B/663